Climate Change and Crop Protection

THE EDITORS

Dr. A.G. Sreenivas M.Sc. (Agri.), Ph.D., PDF (UK & Mexico) is currently is working as Professor of Agricultural Entomology, College of Agriculture, University of Agricultural Sciences (UAS), Raicuhr, Karnataka. He is heading the center for Agro-climatic studies at UAS, Raichur. He has published 113 research papers, 49 research notes, 4 chapters in books and 13 invited lectures.

Dr. G.N. Shreevani, Scientist (Entomology) is currently working at ICAR-Krishi Vigyan Kendra, University of Agricultural Sciences, Raichur. She pursued her Master's and Doctoral degree in Agricultural Entomology at University of Agricultural Sciences, Raichur.

Dr. M.R. Umesh Assistant Professor, University of Agricultural Sciences, Raichur. He earned MS and Ph.D. from UAS Bangalore, specialized in Agronomy and later worked as post doctoral fellow in New Mexico State University, NM USA. He published 59 research articles, 25 book chapters, 12 research bulletins.

Dr. D. S. Aswathanarayana, M.Sc. (Agri.), Ph.D. (Plant Pathology) He is working as Assistant Professor of Plant Pathology in College of Agriculture, UAS, Raichur, Karnataka. He has published 34 national, 11 international research papers and four book chapters.

Dr. J. Ashoka, M.Sc. (Agri.), Ph.D., is working as Professor of Sericulture, College of Agriculture, Raichur and Comptroller, University of Agricultural sciences. He has published 80 research papers, 20 research notes and 2 chapters in books.

Dr. S.G.Hanchinal working as Assistant Professor, University of Agricultural Sciences, Raichur. He earned M. Sc(Agri.) and Ph.D. from UAS Dharwad. He has published 51 research articles, 05 book chapters, 8 research bulletins.

Climate Change and Crop Protection

A.G. Sreenivas
G.N. Shreevani
M.R. Umesh
D.S. Aswathanarayana
J. Ashoka
S.G. Hanchinal

2022

Daya Publishing House®

A Division of

Astral International Pvt. Ltd.

New Delhi – 110 002

Cataloging in Publication Data--DK
Courtesy: D.K. Agencies (P) Ltd. <docinfo@dkagencies.com>

Climate change and crop protection / editors, A.G. Sreenivas [and five others].
pages cm
Contributed articles.
Includes bibliographical references and index.
ISBN 9789354614255

1. Plants, Protection of--India. 2. Crops and climate--India. 3. Climate change mitigation--India. 4. Insect pests--Climatic factors--India. I. Sreenivas, A. G., editor.

LCC S600.64.I4C55 2021 | DDC 630.25150954 23

Published by : **Daya Publishing House®**
A Division of
Astral International Pvt. Ltd.
– ISO 9001:2015 Certified Company –
4736/23, Ansari Road, Darya Ganj
New Delhi-110 002
Ph. 011-43549197, 23278134
E-mail: info@astralint.com
Website: www.astralint.com

Dr. K. N. Kattimani
Hon'ble Chairman & Vice - Chancellor

Foreword

World has become a lot more complicated due rapidly growing populations, demand for food-fibre-feed from all the corners of the globe. Food production and its security in developing countries are always gambled with prevailing weather condition. Agriculture is threatened seriously due to global warming and climate change. Rapid climatic changes are already influencing the availability of critical natural resources which are necessary for the survival of human being. Climate change is not only rise in global temperature and shift in rainfall but also daily weather influence on living organisms. Changes coupled with difficult to manage trends may impair the ability of system to supply basic needs of the present populations. Hyderabad – Karnataka not exceptional in experiencing changes climate and its impact on terrestrial living organisms. Late onset of monsoon, intermittent severe dry spells and terminal water stress are major water resource threat under changing climate. Frequent crop failures, low yield potential, outbreak of certain pests and diseases, dominance of minor pests etc. are also part of the effect of climate change.

In Hyderabad-Karnataka impact of aberrant weather was extreme due to lack bio-diversification, continuous monocropping, intensive paddy cultivation, absence of compatible cropping systems and edaphic factors. Therefore, it becomes necessary to suggest alternate production, protection strategies in a present agriculture scenario of climate change for the region in specific. Develop and understand complete knowledge of source, knowledge of greenhouse gas emission, impact on crops and livestock, mitigation strategies and policy issues is a vital task under climate change. University of Agriculture Sciences, Raichur has also made scientific efforts to study the effect of elevated CO2 and temperature on the major viz., Bt cotton, Redgram, Bengal gram, Maize, Groundnut and sunflower. Most of them showed the positive relation in terms of more growth and biomass but decrease in nutritional status of the crops which aggravates the herbivore damage.

Considering all these issues on consequences of climate change due importance to address the issues and strengthen knowledge compiled publication on climate change and crop protection and agriculture scenario has been brought out. It will also prove to be a very useful reference material for environmentalists, policy makers, scientists, research scholars, extension workers and other scientific community working on climate change. I take this opportunity to congratulate Dr. A.G. Sreenivas and his team for their splendid efforts to compiling, editing and making it available to vast areas scientific fraternity.

(K. N. Kattimani)
Vice-Chancellor
UAS, Raichur

Preface

In recent times, Climatic change and Global warming is one of the major debates worldwide. Climate change is well documented at the global scale andit has been demonstrated through many scientific studies and global data collection that anthropogenic activities have contributed to historically high greenhouse gas levels in the atmosphere. Consequently, there has been a global increase in average temperatures.

Apart from the general effects of climate change, it does have a profound impact on Agriculture's crucial components i.e., the production and protection. Global food production must increase by 50 % to meet the projected demand of the world's population by 2050. Meeting this difficult challenge is even harder if climate change melts portions of the Himalayan glaciers to affect 25 % of world cereal production influencing water availability.Meanwhile, protection of the crops from the ravages of pests and diseases is under the purview of plant protectionists. But, how the climate change will be shaping the food security in coming days is areal question to everybody concerned.

Climate change would have direct and indirect effects on life activities of the crop pests and diseases. It directly affects growth of pest and pathogens; their development, reproduction and survival and indirectly affects by breaking down resistance in temperature sensitive crops. Plant defense systems are lowered under stress in changed climatic conditions, making them more susceptible to the infestation/infection of pests and pathogens. Changes in temperature, rainfall, and wind speed pattern may influence the migratory behaviour of insect pests as well as spread of diseases as well.

We are highly obliged to Dr. K.N. Kattimani, Hon'ble Vice Chancellor, University of Agricultural Sciences, Raichur, Karnataka for writing the foreword of this book. Authors are grateful to our colleague scientists for their contribution in writing the chapters in the book and also giving valuable suggestions in editing the book. This book will be useful for researchers, research scholars, teachers, extension workers and students.

Editors

Contents

1

Climate Change and its Effects on Crop Physiological Processes

M.K. Meena, A.S. Channabasavanna,
M.R. Umesh, and M. Chandra Naik

Introduction

Climate change affects plants growth and development in many different ways. Evaluation of the impacts of climate change on the crop growth and development will be a vital task in developed as well as in under developing countries because of many interdependent crop physiological, physical, biological and chemical processes are ongoing in earth ecosystem. These processes can be affected by change in climate, causing an effect on natural resources (water resources, forest products, *etc.*), on biodiversity, and on plants in general, some positive and on others negative effects, such as, altering biophysical relationship, shrinking of habitats, desertification and general shift in natural world. Increasing in temperature at slow rate *i.e.*, Global warming is directly affects some physiological processes such as rate of plant respiration, photosynthesis, and other biochemical processes. For instance, enhanced CO_2 concentration can increase photosynthetic rate especially for plants growing under warm and dry condition such as C_3 plants. Naturally, plants have their own mechanism to tolerate a certain level of increase in temperature. As soil temperature increases, the decomposition rate of organic matter will increase, and then nutrient mineralization and availability for plants uptake become increased at presence of sufficient water if other conditions are unchanged. Thus, the interaction and different combination effects of rise in CO_2 concentration and temperature is determined by soil properties, water, mineral and nutrient availability *etc.*, as a result, expected response of plants in different environments and climate variability can be either positively or negatively affected.

Therefore, two basic measures are necessary to reduce impacts of climate change on crop plant:- 1) practicing mitigation (reducing causes of climate change) by reducing emission of greenhouse gases (GHGs) from the source, by substitution and conservation of energy, improving carbon sequestration and 2) practicing potential adaptation measures, (*e.g.* reducing the impacts of climate change). Examples such as a) reducing vulnerability (degree of susceptibility of a system to a certain damage) to climate change impacts, focusing on coping strategies and practices to become beneficial by using opportunities associated to climate change by reducing susceptibility and external forces to develop the ability of resilience (increasing tackling capacity of the community and sectors to reduce risk and damages); b) Improving productivity in terms of quality and quantity is vital to satisfy human needs, through adjusting different growth factors and solving effects of extreme events and associated problems, *e.g.* preventing spread of pathogens, weeds, dispersion of insect and pests *etc.*; c) minimize impacts of climate change (its cause and effects) moving forward in researching to identify the responses of plant species to different variable climate conditions, and identifying uncertainty in climate and try to avoid challenges in practicing adaptation and d) Mitigation is an anthropogenic intervention to reduce sources of GHGs or enhance sinks of GHGs (carbon sequestration).

Crop plant is highly sensitive to climate and change in atmosphere, crops having a limited environment in which they are productive and profitable. Henry Nix (1981) of the Australian National University (ANU) posed three questions that need to be addressed when considering what crops can be grown where:

1. For any given crop which areas offer the greatest biophysical advantages?
2. For any given area which crops offer the greatest advantages, and
3. For any given crop or area how may productivity be raised and sustained?

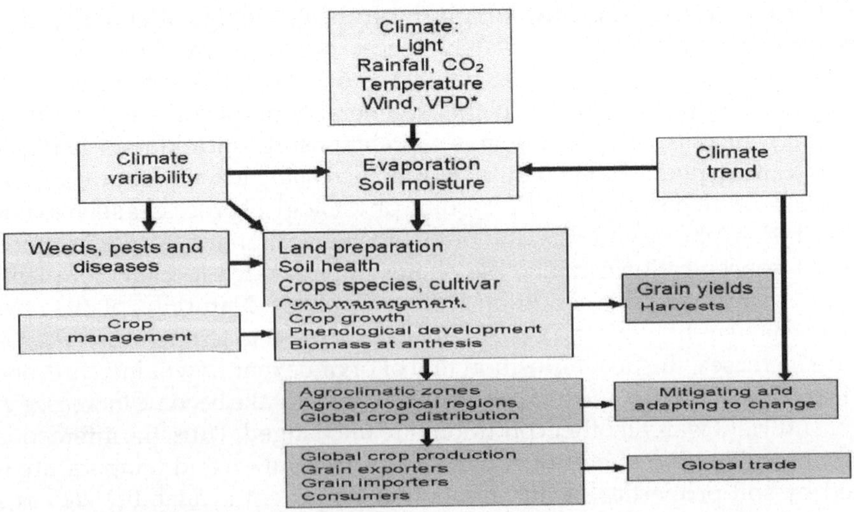

Major Climatic Factors Determinants of Plant Productivity.

Climatic factors *viz.*, light, temperature, CO_2 concentration and relative humidity are fundamental to crop growth. Soil moisture stimulates seeds to germinate, the time to emergence being temperature-dependent. The rate of growth of roots, stem and leaves depends on the rate of photosynthesis, which in turn depends on light, temperature, moisture and carbon dioxide (CO_2), as discussed in detail in the following sections. Temperature and day length also determine when plants produce leaves, stems and flowers, and consequently the filling of grain or the expansion of fruit or seeds of different crop plants. The yield of grain crops depends on grain number and grain weight at harvest, which in turn depends on biomass at anthesis and the availability of moisture post-anthesis. In this section we explore how light, temperature and moisture and other climatic factors determine land use, crop emergence and growth and sale able product.

Significance of Light

Solar radiation is the primary driver of plant photosynthesis, and the radiation levels in the upper atmosphere may be estimated mechanistically as a function of latitude and time of year. Solar radiation at the crop surface differs from incident radiation in the upper atmosphere according to the daily atmospheric transmissivity, determined in part by cloud cover that can vary substantially within and between years

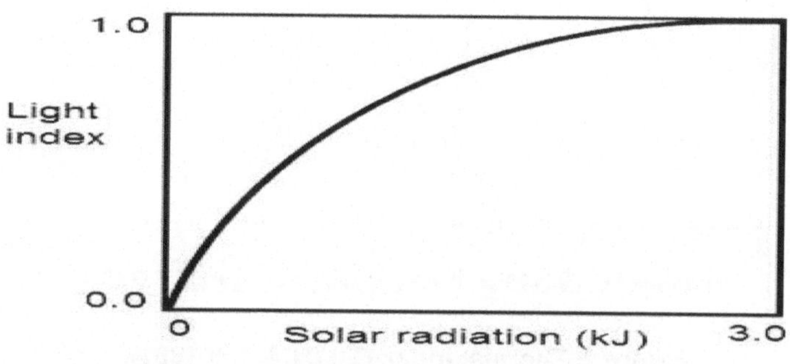

Figure 1: Light index (LI) (H.A. Nix 1981).

Figure 1 shows that the effect of reducing the light reaching the crop canopy on plant growth is most pronounced at low levels of solar radiation. Consequently, in many countries and environments light will often not be the most limiting factor affecting crop growth and yield. However, in mid- to late winter, at very high latitudes or on shaded slopes, or in countries where airborne pollutants cause a substantial reduction in atmospheric transmissivity, then photosynthetic activity and crop yields can be significantly diminished.

Significance of Temperature (°C)

Temperature has major effects on photosynthesis and respiration, plant growth and phenological development. Phenology is particularly important in cooler regions and at higher altitudes. As a rough guide, atmospheric temperatures experienced by crops decrease by about 1°C for each 2°C increase in latitude, or for each 100 m increase in altitude. As shown in figure 2, the bell-shaped responses to temperature shown in figure 3 indicate just how severely growth is limited at low temperatures, whereas at high temperatures the stomata on the leaves can close so as to limit gaseous exchange, water loss, and consequently active growth. These bell-shaped responses to temperature are likely to be distorted by extension on the cool side and truncation on the warm side. Cool temperate climatic plants typically have lower minimum, maximum and optimal temperatures for growth than warm-season C_3 crops and tropical C_4 crops. Temperature is important in controlling phenological changes in development from germination and seedling emergence, through vegetative growth to floral initiation and reproductive growth. Of course variation in temperature tolerance is evident both within populations of single plants and between genotypes. Temperature within a plant, particularly at the growing points, may differ from prevailing air temperatures for a number of reasons, including plant structure and density, distance from the soil surface, and shade. Temperature also has a major influence on the rate of evaporative loss from soils and leaf surfaces as discussed in the following section

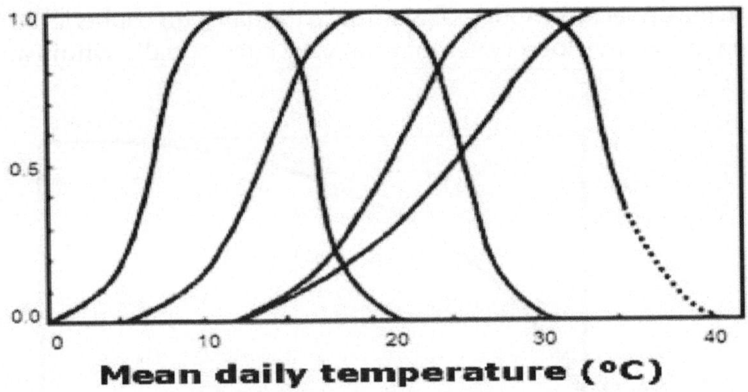

Figure 2: Thermal Index (TI) (H.A. Nix 1981).

Effects of Climate Change on Crop Ecosystems

Many studies particularly on agricultural crops and forest species shows that the enhanced atmospheric CO_2 directly increase biomass productivity, because higher ambient CO_2 concentration stimulates net photosynthetic activity which have been called ´CO_2-fertilization` effect. Transpiration decreases through a partial stomatal closure resulting in increased water use efficiency of plants atleast at a leaf scale; nevertheless, there are considerable differences between different species regarding their response. Some species in terrestrial ecosystems may in the long-term indirectly react negatively, perhaps fatal, to increased CO_2 concentration. The indirect responses of ecosystems are due to the effect of elevated CO_2 concentration is through effect on climate, such as change in temperature or radiation, humidity, precipitation or other climate variables. In most cases, this situation (the change in climate variables) can cause an impact on ecosystems (Bolin *et al.*, 1989).

The uptake of minerals, nutrient and water, canopy exchange of plants, absorption of light energy for the formation of carbohydrate through photosynthesis reactions as well as the breakdown and burning processes of carbohydrate for growth and development of the plant (respiration) is highly dependent on the amount of atmospheric CO_2 concentration and ambient temperature. The processes of transpiration (affected by the opening and closure of stomata) and evaporation from surface of plants determined by the level of temperature and CO_2. Soil biotic processes *e.g.* decomposition, mineralization, immobilization, and soil abiotic processes such as solute transport, weathering, cat ion exchange, *etc.* in the soil are affected by climate change. As a result, it causes a change in net primary production, species composition and resource competition; consequently, the general services, which we get from forest ecosystem; such as forest products, biodiversity, species composition, soil and water resources and recreation are affected (Figure 3).

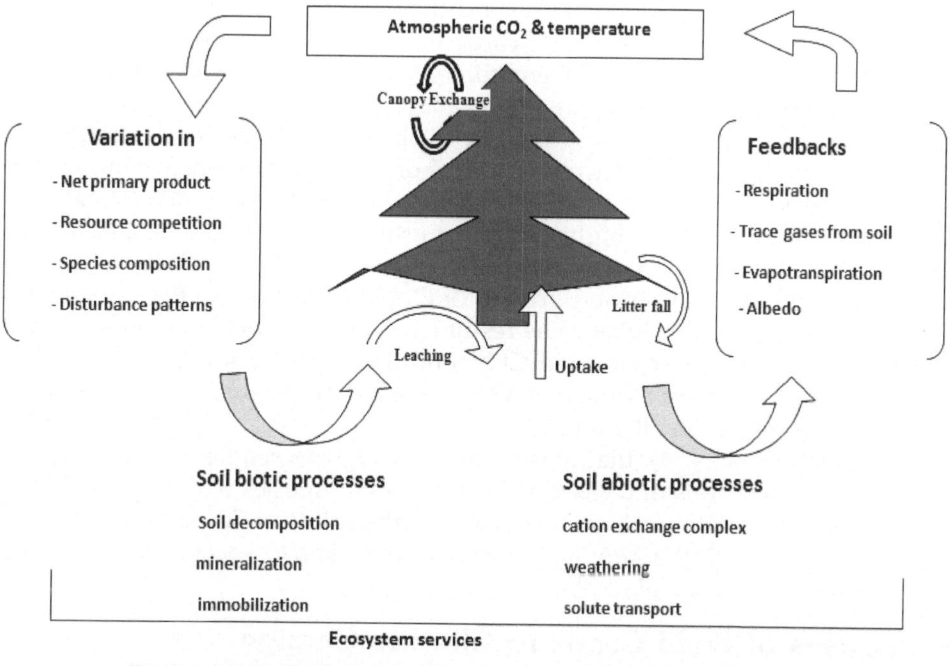

Figure 3: Schematic Diagrams Showing the Most Important Biochemical and Physiological Processes in Plants and Forest Ecosystems Affected by Climate Change (Modified from Lukac *et al.*, 2010).

Rise in Temperature (°C)

Due to climate change the warming of the earth do not uniformly distributed over the world; continents show more rapid temperature increase as compared to oceans. Temperature change will have very different impacts on vegetation and ecosystem productivity, structure and composition depending on the actual temperature range at the location (Morison and Lawrol, 1999). Global temperature has increased about 0.2°C per decade for the past 30 years and warming is larger in the western equatorial pacific than in the eastern equatorial pacific over the past century (IPCC, 2007). Temperature can affect photosynthesis through modulation of the rates of activities of photosynthetic enzymes and the electron transport chain and indirectly through leaf temperatures defining the magnitude of the leaf to air vapor pressure difference, which is a key factor influencing stomata conductance and other plant physiological processes, interlink to each and necessary to proper growth and development.

Crop Plants Responses to Climate Change

Plants are grouped in to C_3, C_4 and CAM plants according to their photosynthetic metabolic pathways. Around 95 % of the world plant biomass grouped in C_3 plant species (*e.g.* wheat, rice, fruits & vegetables), C_4 (*e.g.* maize or corn, sugarcane and sorghum) and CAM (*e.g.* Pineapple and Cactus). These divisions into groups largely based on the enzymes involved in photosynthetic fixation of CO_2, namely Rubisco, PEP carboxylase and to some extent carbonic anhydrase, which are significantly different in their response to CO_2 enrichment. CO_2 together with other minerals can activate Rubisco by binding at a non-catalytic site on the enzyme protein. The process of photorespiration rate is high in C_3 plants and the relative proportion of CO_2 and O_2 inside the leaf determines the rate of photorespiration. In contrast, PEP carboxylase in C_4 plants not inhibited by O_2 and thus photorespiration is negligible. PEP carboxylase also has a higher effective affinity for CO_2 than Rubisco in the absence of O_2. Therefore, we would expect higher atmospheric CO_2 concentrations to increased photosynthesis and growth of C_3 plants but not to the same extent, if any, in C_4 plants (Bolin, 1989). The result from experiments done on wild grass species shows that under elevated CO_2 condition both C_3 and C_4 species show increase in the total plant biomass of 44 % and 33 %, respectively, the increased in C_3 species was greater in tiller formation whereas in C_4 was greater in leaf area. Net CO_2 assimilation rates (*A*), that means (flux of CO_2 between leaf and atmosphere through photosynthesis) increased in both $C_3$and C_4 species with 33 % and 25 %, respectively, while, stomatal conductance (*Gs*) (the ability of CO_2 entering, or water vapor exiting through the stomata) decreased for C_3 and C_4 species by 24 % and 29 %, respectively (Wand *et al.*, 1999).

Responses of Field Crops to Climate Change

Elevated (CO_2 and temperature) leads plants to produce a larger number of mesophyll cell, chloroplasts, longer stems and extended length, diameter and number of large roots, forming good lateral root production with different branching patterns; in some agricultural food crops, resulting in increasing root to shoot ratios under elevated CO_2 and temperature (Qaderi and Reid, 2009). The potential of crop productivity increased under an increased in local average temperature range of 1-3°C, but it decreased above this range (IPCC, 2007), probably the reason could be low vernalization, shortened phenological phases decrease in photosynthesis rate, and increased transpiration. Elevated CO_2 have a positive effect on some annual C_3 field crops, such as soybean, peanut, and rice cultivars, *etc.* Growth and development accelerated throughout the vegetative phase, and before flowering stage started seven days earlier, which contributed to the higher grain yield and change in the chemical composition of the rice grain (Uprety *et al.*, 2010). Some studies also show that a reduction in maize (C_4 species) yield occurred under elevated (CO_2) condition due to shortened growing period and a yield reduction also recorded in some experiment on winter wheat (C_3 species) due to an effect on vernalization period.

Climate Change and Photosynthesis and Plant Respiration Processes in Plants

Respiration can be highly affected by temperature and its rate is determined by status of carbohydrate and supply of adenylate (enzyme catalyzing the conversion processes). The sucrose content of the tissue can govern the capacity of mitochondrial respiration and mitochondrial respiration plays a great role in growth and survival of plants. One would expect atleast a short period increase in respiration rate from parts of plants those show increased growth and assimilation due to elevated CO_2, that is source leaves, individual sink tissue (fruit, seed, steam, root *etc.*) and total sink tissue. Nevertheless, a few reports concluded that long-term treatment with increased concentration of CO_2 resulted in declined whole-plant respiration. Whereas, result of a few other experiments show that a short-term increase in temperature on plants growing in cold climate areas such as Arctic have resulted in greater potential impact on plant respiration than in plants growing in warmer areas (Atkin and Tjoelker, 2003).

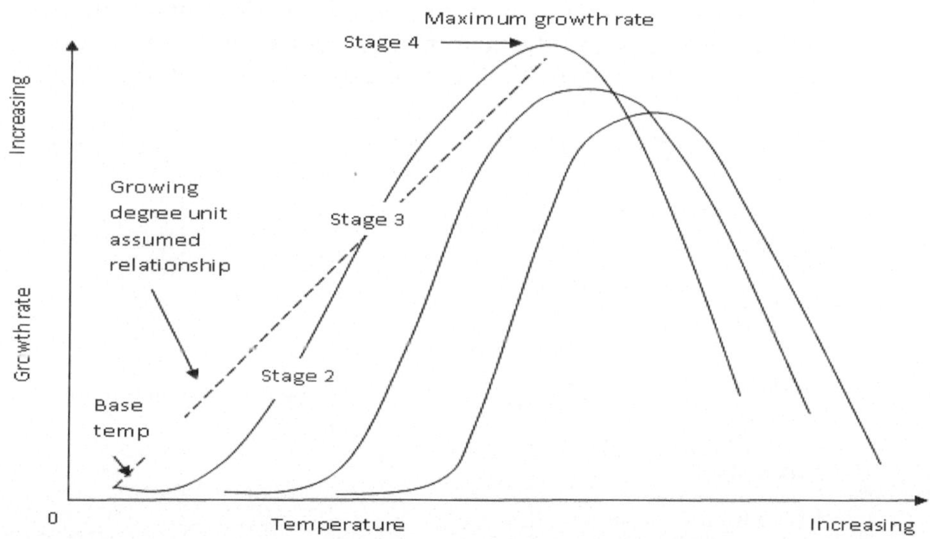

Figure 4(a): Changes in the Rate of Crop Development as a Function of Temperature [*Source*: Modified from Heidorn (2003)]. The three curves in the Figure indicate that optimal temperature range differs and depends on the species and crop type.

Photosynthesis is a process, undertaken in green plant cells to produce sugar molecules, by the uptake of CO_2 from the atmosphere, sun light, energy and water, and during the process oxygen released. Whereas, respiration is a process of oxidation of food or break down of sugars and production of energy for maintenance, reproduction, and growth, which is undertaken in all living organisms at day and night time by the uptake of O_2 and release of CO_2 and energy.

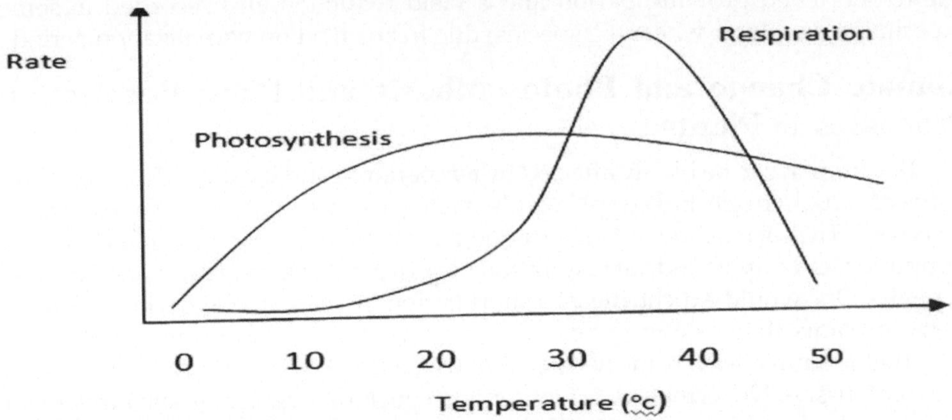

Figure 4(b): Changes in the Rate of Photosynthesis and Respiration of (C_3) Crops as a Function of Temperature (Porter and Semenov, 2005).

Respiration is necessary for many processes in living organisms; for instance, it is crucial for maintenance of photosynthesis activity, mainly because of the energy demands of sucrose synthesis. Moreover, it plays a role in determining the carbon budget of individual plants and the concentration of CO_2 in the atmosphere; it contributes up to 65 % of the total CO_2 released to the atmosphere (Atkin *et al.*, 2005).

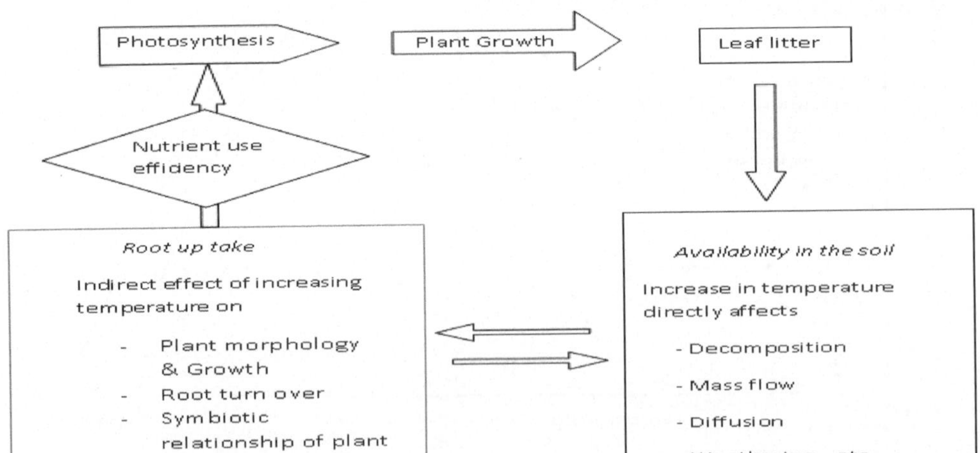

Figure 5: Direct and Indirect Effects of a Rise in Temperature on different Component Processes in Plant and in the Soil (Lukac *et al.*, 2010).

In most plants as temperature increase with optimal range, the rate of respiration as well as the rate of metabolism increase, because increased respiration results with higher energy available, that means as long as nutrients are available, the metabolism processes within the plant will also increase. Following increased temperature to a certain level, the rate of photosynthesis also increases but not as much as respiration. That indicates the amount of CO_2 produced from increased respiration is faster than the amount of O_2 released from increased photosynthesis. Temperature affect photosynthesis through altering the activities of enzymes, electron transport and leaf temperature (leaf-to air vapor pressure difference) can influence the stomatal conductance.

Impacts of Temperature × CO_2 Interaction on Plant Physiological Processes

There are many physiological processes associated with plant growth and development and finally yield of the crop, affected by interaction of both enhanced temperature and carbon dioxide, in processes that determine carbon balance in the shorter term, from the long time scales of development and growth, which together lead to accumulation of biomass and yield. The two main reasons to expect progressively increasing CO_2 responsiveness of plant carbon balance at higher temperatures are 1) the decreased ratio of photosynthesis to photo respiration and 2) the decreased ratio of gross photosynthesis to dark respiration in warmer conditions (Morison and Lawlor, 1999). The effect of elevated CO_2 on photosynthetic reactions are more pronounced in high temperature, *e.g.* around 20°C than at 10°C. Some predictions indicate that future increase in temperature may increase root mortality more in N-rich soils in temperate forests than in N poor soils in boreal forests areas with important implications for the cycling between plant and soil (Lukac *et al.*, 2010). Some (unpublished) studies found that changes in activation state and catalytic constant occur due to both CO_2 and temperature, and there were an interaction, which affected the photosynthetic rate demonstrating the underlying complexity of the photosynthetic regulation mechanisms.

Photosynthetic Responses to CO_2 Concentration

Photosynthetic responses to CO_2 concentration is showing in Figure 6.

Responses of Photosynthesis to Temperature

Photosynthesis can be strongly affected by temperature. Photosynthesis is a biochemical process and its overall temperature response can be understood in terms of the temperature dependencies of its component processes and their well-known. At low to moderate temperatures, the activity of each of these component processes increases with increasing temperature in accordance with the Arrhenius relationship (Farquhar *et al.*, 1980;) At higher temperatures, photosynthesis decreases due to conformational changes in key enzymes. This decrease is reversible at moderately high temperatures but becomes increasingly irreversible with length and intensity of high temperature exposure. Photosynthesis can acclimate considerably to actual growth conditions, but the patterns differ between species (Figure 6). Photosynthesis has a temperature optimum, with the sharpness of the optimum differing between species and growth, and measurement conditions. Optimum temperatures are higher in plants acclimated to higher growth temperatures. In general, as in the

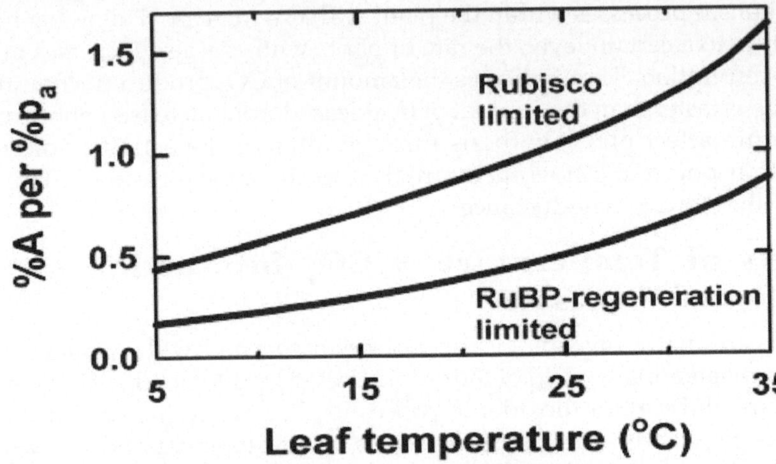

Figure 6: Shows the CO_2 Response of Photosynthesis under different Photosynthetic Limitations.

The response is much greater for Rubisco-limited photosynthesis and at higher temperature. RuBP regeneration-limited photosynthesis responds to CO_2 concentration only because of a changing ratio of carboxylations (that fix CO_2) to oxygenations (that lose CO_2), but the total rate of RuBP regeneration remains the same (Kirschbaum, 1984). In the Rubisco-limited region, on the other hand, the CO_2 uptake kinetics of Rubisco also become a further limitation so that, with increasing CO_2 concentration, there is not only an increasingly favourable ratio of carboxylations to oxygenations, but more RuBP molecules can be utilized as well.

examples in Figure 6, optimum temperatures for photosynthesis acclimate by about 0.5°C per 1.0°C change in effective growth temperature (Battaglia *et al.*, 1996). *Atriplex lentiformis, Atriplex sabulosa,* and *Tidestromia oblongifolia* are all C_4 plants whereas *Larrea divaricata* is a C_3 plant. This indicates that both C_3 and C_4 species can be capable of high photosynthetic rates up to extremely high temperatures. While C_3 plants are more typically found at cooler locations than C_4 plants (Figure 7), the differences in distribution are not necessarily reflected in the photosynthetic response curves of individual species. Photosynthetic responses to temperature are thus highly dependent on species and growth conditions.

Conclusion

Industrialization and unmanaged utilization of resources results in higher rate of GHG emissions and accumulations in the atmosphere, which might result in effects many years later. In addition, population growth, urbanization and industrialization will result in a higher demand of renewable and non-renewable natural resources. Even though, currently there is a gap between availability (supply) and demand of increased population interests, there is a need to consider a more careful utilization of natural resources. Activities of filling the gap must be by reducing emission and protecting the coming generation, and make them beneficial. Crops are dependent on light, temperature, moisture and carbon dioxide (CO_2) concentration to produce the grains and other crop products that are so essential to our nutrition, health and

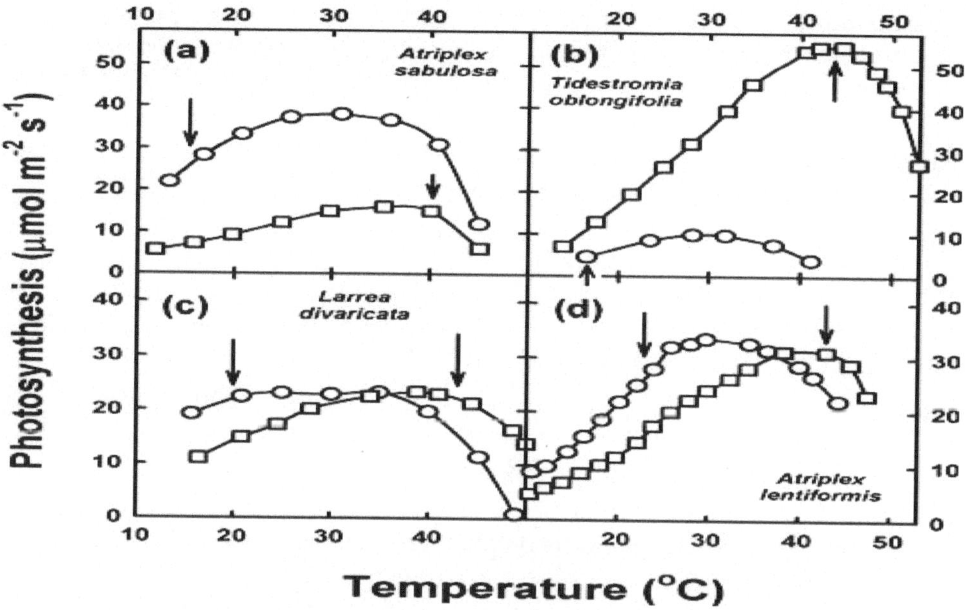

Figure 7: Response of C_3 and C_4 Plants to Temperature.

well-being. However, the levels of these climate inputs, particularly rainfall, vary between locations and years, in part due to climate variability. Temperature and water supply also vary over the long-term, including in response to climate change, with major implications for crop production, and the livelihood of crop producers. Thereby reduce impacts of climate change and minimize adaptation challenges. Forest trees, through photosynthesis, remove CO_2 from the atmosphere and store it as organic carbon during life of the tree, but when the tree dies and decay or being burnt most of the carbon will be released back to the atmosphere as CO_2.

References

Atkin, O. K. and Tjoelker, M. G., 2003. Thermal acclimation and the dynamic response of plants respirationto temperature: Trends in Plant Sciences., 8(7)., TX 77843-2135.

Atkin, O. K., Bruhn, D., Hurry, V. M. and Tjoelker, M. G., 2005. The hot and the cold: unraveling the variable response of plant respiration to temperature: Functional Plant Biology, 32, 87-105.

Battaglia, M., Beadle, C. and Loughhead, S., 1996. Photosynthetic temperature response of Eucalyptus globulus and Eucalyptus nitens. Tree Physiology 16, 81-89.

Bolin B., Döös B. R., Jager J. and Warrick R. A., eds., 1989. The Green House Effect climate change and Ecosystems: SCOPE 29. International Council of Scientific Union. Great Britain.

Farquhar, G. D., Von Caemmerer, S., and Berry, J. 1980. A biochemical model of photosynthetic CO_2 assimilation in leaves of C_3 species. Planta., 149: 78-90.

Kirschbaum, M. U. F. and Farquhar (1984). The sensitivity of C_3 photosynthesis to increasing CO_2 concentration. A theoretical analysis of its dependence on temperature and background CO_2 concentration. Plant, Cell and Environment., 17: 747-754.

Heidorn, K. C., 2003. The weather doctor's, weather almanac growing by degrees for.http://www.islandnet.com

IPCC, 2007: Summary for Policymakers. In: Climate Change 2007: Mitigation. Contribution of Working Group III to the Fourth Assessment Report of the Intergovernmental Panel on Climate Change [B. Metz, O.R. Davidson, P.R. Bosch, R. Dave, L.A. Meyer (eds)], Cambridge University Press, Cambridge, United Kingdom and New York, NY, USA.

Lukac, M., Calfapietra, C., Lagomarsino, A. and Loreto, F., 2010. Global climate change and tree nutrition: effects of elevated CO_2 and temperature: Tree physiology [0829-318X] Lukac YR. 30, 1209 -1220.

Morison, J.I.L, and Lawlor D.W., 1999. Interaction between increase CO_2 concentration and temperature on plant growth. Plant Cell and Environment., 22: 659-682.

Nix, H.A., 1981. Simplified simulation models based on specified minimum data sets: the cropeval concept. In Application of remote sensing to agricultural production forecasting, edited by A. Berg, AA Balkema, Rotterdam, pp. 151-169.

Porter, J.R., and Smenenov, M.A., 2005. Crop responses to climatic variation: Phil. Trans. R.Soc.B 360, 2021- 2035.

Qaderi, M. M. and Reid D. M., 2009. Crop responses to elevated carbon dioxide and temperature, In Singh S. N., (ed.), Climate Change and Crops, Environmental Science and Engineering, DOI 10.1007/978-3-540-88246-6 1, Springer-Verlag Berlin Heidelberg.

Uprety D. C., Sen S., and Dwivedi N., 2010. Rising atmospheric carbon dioxide on grain quality in crop plants: physiol Mol Biol Plants.

Wand, S. E., Midgley, G. F., Jones, M. C., and Curtis, P. C., 1999. Responses of wild C_4 and C_3 grass (Poaceae) species to elevated atmospheric CO_2 concentration: a meta-analytic test of current theories and perceptions: Global Change Biology 5, 723-741.

2

Effect of Climate Change on Dynamics of Insect Pests

A.G. Sreenivas and G.N. Shreevani

Introduction

Despite tremendous improvements in technology and crop yield potential, food production remains highly dependent on climate since solar radiation, temperature, and precipitation are the main drivers of crop growth. Meanwhile, plant diseases and pest infestations, as well as the supply of and demand for irrigation water are also influenced by climate. The increase of temperature and atmospheric CO_2 have been very significant during the last three decades (Stern, 2007) affecting all sectors of agriculture. It has been reported by federal agencies that, CO_2 concentration has been increased by approximately 30 per cent since the industrial revolution which is believed to be responsible for an increase of about 0.66°C in mean annual global surface temperature. Global climate models have projected the effects of increasing atmospheric concentrations of greenhouse gases. Recent simulations predict a mean global warming of 1.5 to 5.8 °C (2.7 to 10.4 °F) by the end of the century for varying scenarios of population growth, economic development, energy use, and land-use change (IPCC, 2000 and 2001). Because a warmer atmosphere can hold more water vapor, they also predict an increase in mean global precipitation of 5 to 15 %. The atmospheric CO_2, which is considered to be chiefly responsible for the greenhouse effect, has increased from approximately 310 ppm in 1950 to about 400 ppm in the year 2011. This concentration is estimated to reach levels of 421 to 936 ppm by the end of the 21st century, depending on the magnitude of future human activities (IPCC, 2013).

Climate change can result in expansion of deserts, reduced glaciers, ice and snow, severe droughts, heavy rains, severe heat waves, ocean acidification, rise in sea levels, disturbed agricultural systems leading to reduced food production and loss of habitats, species extinction and loss of diversity. Extreme weather events, which occur in every agricultural region of the world, cause severe crop and livestock damage. The extent of crop damage depends on the duration of stress and crop developmental stage. Crop yields are most likely to suffer if the adverse weather conditions, especially high temperature and excess or deficit precipitation occur during critical developmental stages such as the early stages of plant reproduction.

Within agriculture, how the climate change impacts insect pests and diseases is an important research area that is engaging biological scientists. These climatic and weather changes not only affect the status of insect pests but also affect their population dynamics, distribution, abundance, intensity and feeding behaviour. Intensity of change in climatic ecosystem noted by meteorological science has showed direct and also indirect effects on the prey-host relationship, their immune responses and rate of development, their fecundity and various physiological functions.

Climate Change and Crop Yields

Increased Green House Gases (GHGs) and the consequent global warming has resulted in climate change including change in the amount and pattern of precipitation, more restricted water availability, more frequent occurrence of extreme weather events including heat waves, droughts and heavy rainfall (Battisti and Naylor, 2009). These changes in climate are expected to have profound effects on world agriculture, because any changes in weather will influence plant growth and increase uncertainty regarding food production due to decreasing crop yields and the loss of cropping areas. Along with these changes, will be longer growing seasons and possible change in geographical regions suitable for crop production. There is considerable debate about the effects of global warming on agricultural production. There has been a tendency in the literature towards the view that agricultural damages over the next century will be minimal, and that a few degrees Celsius of global warming would be beneficial for agriculture (Cline, 2007).

A global assessment of the potential impact of climate change on world food supply by Rosenzweig *et al.* (1993) suggests that "doubling of the atmospheric carbon dioxide concentration will lead to only a small decrease in global crop production." Investigations regarding the effect of global climate change on crop yield have proposed substantial losses in productivity under the global surface temperature increases projected to result by the end of the twenty first century (Reddy *et al.*, 2002).

Climate Change and Pest Dynamics

Climatic factors like temperature and precipitation in particular, have a very strong influence on the development, reproduction and survival of insect pests. Researchers found that the numbers of leaf eating insects are likely to surge as a result of rising levels of CO_2, at a time when crop production will have to be boosted to feed an extra three billion people living at the end of 21st century (Connor, 2008). It is predicted that some extreme events will increase in frequency as a result of a change in natural climate variability (McCarthy *et al.*, 2001). Such changes in climatic conditions could profoundly affect the population dynamics and the status of insect pests of the crops (Woiwood, 1997). These effects could either be direct, through the influence of weather on the insects' physiology and behaviour (Samways, 2005; Parmesan, 2007 and Merrill *et al.*, 2008), or may be mediated by host plants, competitors or natural enemies (Harrington *et al.*, 2001 and Bale *et al.*, 2002).

Climate change related factors like rise in temperature, changes in precipitation patterns, milder and shorter winters, rise of sea levels and increased incidence of extreme weather events can directly influence insects by affecting their rate of development, reproduction, distribution, migration and adaptation. In addition, indirect effects can occur through the influence of climate on the insect's host plants, natural enemies and interspecific interactions with other insects. The impacts include changes in phenology, distribution and community composition of ecosystem that finally leads to extinction of species (Walther *et al.*, 2002). As insects represent huge numbers of taxa and individuals, with their short generation times, high mobility and high reproductive rates, they will respond more quickly to climate changes than longlived organisms, like higher plants and mammals (Menendez, 2007). Infact, insects may be the first predictors of climate change. Milder and shorter winters will result in early start by the pest under warm weather condition breeding (Bale *et al.*, 2002). Other changes include expanded pest ranges, disruption of synchrony between pests and natural enemies and increased frequency of pest outbreaks and upheavals (Parmesan, 2007 and Van Asch and Visser, 2007).

Rising Temperature and Pest Population

A key factor regulating the life history pattern of insect pest is temperature. Because insects are poikilothermic (cold-blooded) organisms, the temperature of their bodies is approximately the same as that of the environment. Therefore, the developmental rates of their life stages are strongly dependent on temperature. Almost all the insects will be affected to some degrees by changes in temperature and there may be multiple effects upon insect life histories. Laboratory and 303 modeling experiments support the notion that the biology of agricultural pests are likely to respond to increased temperatures (Fye and McAda, 1972, Cammell and Knight, 1991 and Fleming and Volney, 1995). With every degree rise in global temperature, the life cycle of insect will be shorter. The quicker the life cycle, the higher will be the population of pests. In temperate regions, most insects have their growth period during the warmer part of the year because of which, species whose niche space is defined by climatic regime, will respond more predictably to climate change while those in which the niche is limited by other abiotic or biotic factors will be less predictable (Bale *et al.*, 2002). In the first case, the general prediction is that if global temperatures increase, the species will shift their geographical ranges closer to the poles or to higher elevations and increase their population size (Sutherst, 2000, Harrington *et al.*, 2001, Bale *et al.*, 2002 and Samways, 2005). The increase in temperature associated with climatic change, would impact crop pest insect populations in several complex ways like (a) extension of geographical range (b) increased over-wintering (c) changes in population growth rate (d) increased number of generations (e) extension of development season (f) changes in crop pest synchrony (g) changes in inte-rspecific interactions (h) increased risks of invasions by migrant pests and (i) introduction of alternative hosts and overwintering hosts. But all these effects of temperature on insects largely overwhelm the effects of other environmental factors (Bale *et al.*, 2002). With the increase in mean temperature by ~0.6°C over the past century and projected increases in the future (IPCC, 2001), climate change has been shown to have effects on ecosystems worldwide (Walther *et al.*, 2002). Temperature increases already have caused changes in species diversity and distribution. It will alter the distribution of many species in different taxa (Hickling *et al.*, 2005). It has been recognized that global warming affects the individual species and communities in the form of range shifts and extinctions (Walther *et al.*, 2002, Root *et al.*, 2003 and Battisti, 2004). Depending on the development strategy of an insect species, temperature can exert different effects (Bale *et al.*, 2002). Temperature can impact insect physiology and development directly or indirectly through the physiology or existence of hosts. Some insects take several years to complete one life cycle.

These insects (cicadas, arctic moths) will tend to moderate temperature variability over the course of their life history. Some crop pests are "stop and go" developers in relation to temperature, so they develop more rapidly during periods with suitable temperatures. It has been estimated that with a 2°C temperature increase, insects might experience one to five additional life cycles per season (Yamamura and Kiritani, 1998). Warming could decrease the occurrence of severe cold events, which could in turn expand the over-wintering area for 304 insect pests (Patterson *et al.*, 1999). In-season effects of warming include the potential for increased levels of feeding and growth, including the possibility of additional generations in a given year (Cannon, 1998). Migratory insects may arrive earlier or the area in which they are able to over-winter may be expanded. Natural enemy and host insect populations may respond differently to changes in temperature. Parasitism could be reduced if host populations emerge and pass through vulnerable life stages before parasitoids emerge. Hosts may pass though vulnerable life stages more quickly at higher temperatures, reducing the window of opportunity for parasitism. Temperature may change gender ratios of some pest species such as thrips (Lewis, 1997) potentially affecting reproduction rates. However, insects that spend important parts of their life history in the soil may be more gradually affected by temperature changes than those that are above ground.

This is because soil provides an insulating medium that will tend to buffer temperature changes more than the air (Bale *et al.*, 2002). Lower winter mortality of insects due to warmer winter temperatures could be important in increasing insect populations (Harrington *et al.*, 2001). At higher temperatures, aphids have been shown to be less responsive to the aphid alarm pheromone they release when under attack by insect predators and parasitoids, resulting in the potential for greater predation (Awmack *et al.*, 1997). Increases in mean temperatures particularly, milder winters and longer summers, are highly favorable to increased aphid populations and are thought to have caused extensions to the geographical range of many insect pests, leading to increased range and severity of infestations. There is also the prospect of new pests, which may become much more important as a result of increased temperature due to global warming.

Elevated CO_2 and Pest Population

One of the most studied aspects of climate change is the effect of increasing concentrations of CO_2 on plants. Plants consist primarily of carbon and elevated CO_2 levels allow them to grow more rapidly because they can assimilate carbon more quickly. Greenhouse growers have known this for decades and add CO_2 to encourage plant growth. Similarly, because CO_2 increases the photosynthetic rates of most crop plants, scientists initially thought that increasing CO_2 would be a solution for the world's food supply (LaMarche *et al.*, 1984). In addition to enhanced growth, many crop plants become more drought-tolerant due to CO_2 enrichment. This is because the openings in the leaves (stomata) that let CO_2 in also let water vapor out and if there is high CO_2 concentration in the vicinity of leaf then the stomata need not open as much. It was suggested that under conditions of elevated CO_2, plants will produce better yields even when conditions are harsh (LaMarche *et al.*, 1984). Unfortunately, these optimistic predictions have not proven accurate. One reason for this is that insects also eat more when plants are grown under elevated levels of CO_2 to compensate their low nutritional quality. A rise in CO_2 generally increases the carbon to nitrogen ratio of plant tissues thereby reducing the nutritional quality for protein limited insects diluting the nitrogen content of the tissues (Coviella *et al.*, 1999). The expected reactions from herbivores to the increase in carbon to nitrogen ratio are compensatory feeding, concentrations of defensive chemicals in plants and competition between pest species. Insects may accelerate their food intake to compensate for reduced leaf nitrogen content (Holton *et al.*, 2003), although this is not always the case (Knepp *et al.*, 2005). However, the response of plants to increased CO_2 varies among species. Increased carbon to nitrogen ratios in plant tissue may slow insect development and increase the length of life stages vulnerable to attack by parasitoids. Phytophagous insects may also develop adaptations to overcome higher carbon to nitrogen ratios, for example the pine sawfly *Neodiprion lecontei*, showed an increase in the efficiency of nitrogen utilization when reared on plants treated with high CO_2 concentration (Williams *et al.*, 1994). However, other insect species seem unable to compensate the lower nutritional quality of the plants by increasing the efficiency of nutrient utilization (Brooks and Whittokar, 1999). The experiments of Lindroth *et al.* (1993) on three species of saturnid moths showed that the performance of caterpillars is only marginally affected when the nitrogen content of the leaves is reduced by 23 % and the carbon to nitrogen ratio increased by 13-28 %.

Conclusions

Insects are a recognized threat to agricultural production throughout the world. Most insects can adapt their body temperature to the temperature of the environment. Many fear that global warming will affect insects' metabolism, allowing them to increase their multiplication rate. Rising temperatures will open new areas for colonization by insects and more of them will spread to newer areas. Increase in the populations of some insects may also take place as a result of higher multiplication rates, along with the elimination for the need to go into diapauses during winter to avoid colder temperatures. The effects could be further amplified under conditions where alternate host plants are already available for wintering.

References

Awmack, C. S., Woodcock, C. M. and Harrington, R., 1997, Climate change may increase vulnerability of aphids to natural enemies. Ecol. Entomol. 22: 366-368.

Bale, J., Masters, G., Hodkinson, I., Awmack, C., Bezemer, T., Brown, V., Butterfield, J., Buse, A., Coulson, J. and Farrar, J., 2002. Herbivory in global climate change research: direct effects of rising temperature on insect herbivores. Global Change Biol. 8: 1-16.

Battisti, A., 2004, Forests and climate change-lessons from insects. Forest. 1: 17-24.

Battisti, D.S. and R.L. Naylor.2009.Historical warnings of future food insecurity with unprecedented Seasonal heat. Science., 323: 240-244.

Brooks, G.L. and Whittaker, J.B., 1999. Responses of three generations of a xylem-feeding insect, Neophilaenus lineatus (Homoptera), to elevated CO2. Global Change Biol. 5: 395-401.

Cammell, M.E.and Knight J.D. 1991. Effects of climate change on the population dynamics of Crop pests. Adv. Ecol. Res. 22: 117- 62.

Cannon, R.J., 1998, The implications of predicted climate change for insect pests in the UK, with emphasis on non-indigenous species. Global Change Biol. 4: 785-96.

Cline, W. R. 2007. Global Warming and Agriculture. Impact Estimates by Country. Center for Global Development, Peterson Institute for International Economics., Washington, DC.

Connor, D.J., 2008. Organic agriculture cannot feed the world. Field Crops Res. 106: 187-190.Coviella, C. and Trumble, J., 1999, Effects of elevated atmospheric carbon dioxide on insect plant interactions. Conserv. Biol. 13: 700-712.

Coviella, C.E, Stipanovic, R.D. and Trumble, J.T., 2002. Plant allocation to defensive compounds: interactions between elevated CO2 and nitrogen in transgenic cotton plants. J. Exper. Bot. 53: 323-331.

Fye, R.E., Mc Ada, W.C., 1972. Laboratory studies on the Development, Longevity, and Fecundity of Six Lepidopterous Pests of Cotton in Arizona (Washington, DC: USDA leming, R.A.and Volney, W.J., 1995. Effects of climate change on insect defoliator population processes in Canada's boreal forests: some plausible scenarios. Water Air Soil Pollution 82: 445-54.

Harrington, R., Fleming, R. and Woiwod, I., 2001. Climate change impacts on insect management and conservation in temperate regions: can they be predicted? Agric. For. Entomol., 3: 233-240.

Hickling, R., Roy, D. B., Hill, J.K. and Thomas, C.D., 2005. A northward shift of range margins in British Odonata. Global Change Biol. 11: 502-506.

IPCC (Intergovernmental Panel on Climate Change), 2001, Climate Change 2001: The Scientific Basis-Contribution of Working Group 1 to the third assessment report of the IPCC. Cambridge.

IPCC (WGI). 2000. Special Report on Emissions Scenarios. Cambridge: Cambridge University Press.

IPCC (WGI). 2001. Climate Change 2001: The Scientific Basis. Contribution of Working Group I to the Third Assessment Report of the Intergovernmental Panel on Climate Change. J.T. Houghton and Ding Yihui, eds. Cambridge: Cambridge University Press.

IPCC, 2013, Climate Change 2013: The physical science Basis. Contribution of working group I to the fifth assessment report of the IPCC, Ed. Stocker, T. F., Qin, D., Plattner, G. K., Tignor, M., Allen, S. K., Boschung, J., Nauels, A., Xia, Y., Bex, V. and Midgley, P. M., Cambridge University press, Cambridge, United Kingdom and New York, NY, USA, p. 1535.

Knepp, R.G., Hamilton, J. G., Mohan, J.E., Zangerl, A.R., Berenbaum, M.R. and DeLucia, E.H., 2005. Elevated CO2 reduces leaf damage by insect herbivores in a forest community. New Phytologist. 167: 207-218.

LaMarche, V.C., Grabyll, D.A., Fritts, H.C. and Rose, M.R., 1984. Increasing atmospheric carbon dioxide: Tree ring evidence for growth enhancement in natural vegetation. Sci. 22: 5-21.

Lewis, T., 1997, Thrips as crop pests. CAB International, Oxon, GB.

Lindroth, R.L., Kinney, K.K. and Platz, C.L., 1993. Responses of deciduous trees to elevated atmospheric CO2: productivity, phytochemistry and insect performance. Ecol. 74: 763-777.

McCarthy, J.J., Canziani, O.F., Leary, N.A., Dokken, D.J. and White KS. 2001. Climate change 2001: Impacts, Adaptation, and Vulnerability. United Nations Intergovernmental Panel on Climate Change. Available at www.grida.no/climate/ipcc tar/wg2/001htm.

Menendez, R., 2007. How are insects responding to global warming? Tijdschrift voor Entomologie. 150: 355-365.

Merrill, R., Gutie´rrez, D., Lewis, O., Gutie´rrez, J., Diez, S. and Wilson, R., 2008. Combined effects of climate and biotic interactions on the elevational range of a phytophagous insect. J. Animal Ecol. 77: 145-155.

Parmesan, C., 2007. Influences of species, latitudes and methodologies on estimates of phenological response to global warming. Global Change Biol. 13: 1860-1872.

Patterson, D.T., Westbrook, J. K., Joyce, R. J. V., Lingren, P.D. and Rogasik, J., 1999, Weeds, insects and diseases. Climate Change. 43: 711-27.

Root, T. L., Price, J.T., Hall, K. R., Schneider, S. H., Rosenzweig, C. and Pounds, J. A., 2003, Fingerprints of global warming on wild animals and plants. Nature., 421: 57-60.

Rosenzweig, C., Parry, M. L., Fischer, G. and Frohberg, K., 1993. Climate change and world food supply. Research Report No. 3. Oxford: University of Oxford, Environmental Change Unit.

Samways, M., 2005. Insect Diversity Conservation. Cambridge University Press, Cambridge, 342.

Stern, N., 2007. The economics of climate change: the stern review. Cambridge University Press, Cambridge. doi: 10.1017/CB09780511817434.

Van Asch, M. and Visser, M.E., 2007. Phenology of forest caterpillars and their host trees: the importance of synchrony. Ann. Rev. Entomol. 52: 37-55.

Walther, G.R., Post, E., Convey, P., Menzel, A., Parmesan, C. and Beebee, T. J. C., 2002. Ecological responses to recent climate change. Nature. 416: 389-395.

Williams, R.S., Lincoln, D.E. and Thomas, R.B., 1994. Loblolly pine grown under elevated CO2 affects early instar pine sawfly performance. Oecologia. 98: 64-71.

Woiwod, I., 1997. Detecting the effects of climate change on Lepidoptera. J. Insect Conser. 1: 149-158.

Yamamura, K. and Kiritani, K., 1998, A simple method to estimate the potential increase in the number of generations under global warming in temperate zones. Appl. Entomol. Zoology. 33: 289- 298.

3

Emerging Pest vis-a-vis Climate Change

S.G. Hanchinal, A.G. Sreenivas and G. Akshatha

Introduction

Insect pests cause an estimated annual loss of 10.8% globally, and the extent of losses in India has been estimated to be 15.7% (Dhaliwal *et al.,* 2015). The pest associated losses likely to increase as a result of changes in crop diversity and climate change, and changes in the cropping patterns and the cropping intensity (Table 1). Pest associated losses increased from an average of 7.2 % during the pre-green revolution period to 23.3 % during the post green revolution period in different crops in India. This is in part because of the unpredictable adaptation of certain pests to a changing environment which creates situations of uncertainty (Chakraborty and Newton 2011).

Insects are likely to be affected by climate change because they are ectothermic and sensitive to temperature (Bale *et al.,* 2002). Effects of climate change on insect pests is of greater importance as the insects are involved in many biotic interactions such as plants, natural enemies, pollinators and other organisms, which are the key players for the ecological functioning. The effect can be direct, through the influence of climatic factors on the insects' physiology and behaviour (Parmesan 2007) or indirect, as mediated by host plants, competitors or natural enemies (Lastvka 2009; Thomson *et al.,* 2010). As a result of global climate change, pest populations may become unstable, leading to outbreaks in some areas, causing greater losses; while in others, losses due to pests may decline. Climate change provides suitable conditions for such pests to adapt across the areas which were previously detrimental for their survival (Chakraborty 2013; Fletcher *et al.,* 2010).

Indeed, the main consequence of climate change and accelerated globalization is a heightened level of unpredictability regarding spatial and temporal interactions between weather, cropping systems, and pests (Chakraborty, 2013). This is the ongoing crop protection challenge that farmers, advisers, researchers and policy makers face.

Climate Change: Possible Effects on Insect Pests

Effect of Increasing Atmospheric Carbon Dioxide on Plant Insect Interaction

A larger crop canopy and denser foliage resulted from enhanced CO_2 level in the atmosphere will create more relative humidity, thereby making micro-environment more favorable to pests. Increases in food quality, *i.e.* increase in the nitrogen content of plants due to high temperature, can results in sudden resurgence of population of pests. Moreover under condition of stress, plant defensive systems are less effective and they become more susceptible to pest attack (Coviella and Trumble 1999). Some more findings on effects of enhanced CO_2 on crop pests are given in the (Table 2).

Table 1: Pest Associated Losses in different Crops during the Pre- and Post-Green Revolution in India

Crop	*Pre-green Revolution (Early 1960s)*	*Post-green Revolution (Early 2000s)*	*Changes in Pest Associated Losses*
Cotton	18	50	+32
Groundnut	5	15	+10
Other oilseeds	5	25	+20
Pulses	5	15	+10
Rice	10	25	+15
Maize	5	25	+20
Sorghum and millets	3.5	30	+26
Wheat	3	5	+2
Sugarcane	10	20	+10
Average	7.2	23.3	+16.1

Source: Dhaliwal *et al.* (2010).

Table 2: Effect of Increasing Atmospheric Carbon Dioxide on Plant Insect Interaction

Increasing Atmospheric Carbon Dioxide Leads to Increasing	*Reference*
Food consumption by caterpillars	Osbrink *et al.*, 1987
Reproduction of aphids	Bezemer *et al.*, 1999
Predation by lady beetle	Chen *et al.*, 2005
Carbon based plant defences	Coviella and Trumble, 1999
Effect of foliar application and *B.thuringiensis*	Coviella and Trumble, 2000

Increasing Atmospheric Carbon Dioxide Leads to Increasing	***Reference***
Insect development rates	Osbrink *et al.*, 1987
Response to alarm pheromones by aphids	Awmarck *et al.*, 2000
Parasitism	Roth and Lndroth, 1995
Effect of transgenic *B. thuringiensis*	Coviella *et al.*, 2000
Nitrogen-based plant defense	Coviella and Trumble, 1999

Impact of Enhanced Temperature on Crop Pest Dynamics

Insects are cold-blooded organisms - the temperature of their bodies is approximately the same as that of the environment. Therefore, temperature is probably the single most important environmental factor influencing insect behaviour, distribution, development, survival, and reproduction. Exposure to temperatures on either side of the range ('threshold of development or 'developmental zero' exerts an adverse impact on the insect by slowing down the speed of development. It has been estimated that with a CO_2 temperature increase, insects might experience one to five additional life cycles per season (Yamamura and Kiritani, 1998).

Other researchers have found that moisture and CO_2 effects on insects can be potentially important considerations in a global climate change setting (Coviella and Trumble 1999; Hunter 2001). Rising temperatures are likely to result in the availability of new pests and pest niches. Insects respond to climate change in various ways, ranging from changes in phenology and distribution to influencing community dynamics and composition. Furthermore, specialist insect pests inhibiting narrow niches in extreme environments are more likely to be adversely affected than the polyphagous insect pests that show phenotypic and genotypic plasticity and occur across a range of environments (Moore and Allard, 2008).

Most analyses conclude that in a changing climate, pests may become even more active than they currently are, thus posing the threat of greater economic losses to farmers (Coakley *et al.*, 1999). Some crop pests are "stop and go" developers in relation to temperature. They develop more rapidly during periods of time with suitable temperatures. Increased temperatures will accelerate the development of these types of insects-possibly resulting in more cycles of generations (and crop damage) per year. Some more findings are given in Table 3.

Table 3: Effect of Enhanced Atmospheric Temperature on Crop Pest Dynamics

***Increasing Atmospheric Temperature Leads to* Increasing**	***Reference***
Northward migration	Parmesan, 2006
Migration up elevation gradients	Epstein *et al.*, 1998
Insect developmental rates and oviposition	Regniere, 1983
Potential for insect outbreaks	Bale *et al.*, 2002
Invasive species introductions	Dukes and Mooney, 1999
Insect extinctions	Thomas *et al.*, 2004

***Increasing Atmospheric Temperature Leads to* Increasing**	***Reference***
Effectiveness of insect biocontrol by fungi	Stacy and Fellowes, 2002
Reliability of economic threshold levels	Stacy and Fellowes, 2002
Insect diversity in ecosystems	Erasmus *et al.*, 2002
Parasitism	Hance *et al.*, 2007

Source: D.K. Das *et al.*, 2011.

Climate Change: Possible Effects on Emerging Pest Scenario or Pest Evolution under Climate Change

Several insect pests, that were important in the past or the minor pests, are likely to become more devastating with global warming and climate change (Sharma, 2013) (Figure 1, Table 4). Many insect species, that will move to newer areas as invasive pests will also pose a major threat to crop production and food security, as they find more suitable climatic niches in the new areas. The invasive species are likely to cause more harm in the absence of natural enemies in the new habitats.

Figure 1: Pest Outbreaks Due to Climate Change (a = Pod borer, *Helicoverpa armigera* damage in pigeon pea following wet weather conditions in Sept – Oct, b = Mealy bug, *Ceroplastodes cajaninae* infestation in pigeon pea under prolonged hot and dry conditions, c = Beet armyworm, *Spodoptera exigua* damage in chickpea triggered by winter rains on Oct – Nov, and d = Pink stem borer, *Sesamia inferens* damage in sorghum due to hot and dry conditions during the post rainy season).

Climate change can lead to higher densities of *Helicoverpa armigera* (Hub.) and *Maruca vitrata* (Fab.) populations and the shifting to the temperate regions, which will have drastic effect on legumes and related crops. Occurrence of *H. armigera* as an invasive pest in Brazil and North America has been attributed to the climate change. The woolly aphid, Ceratovacuna lanigera Zehntner has emerged as a serious pest in Maharashtra due to climate change. Earlier reported in West Bengal in 1958, it has now invaded Central and South India. Further, planting dates corresponding to different temperatures and relative humidity also affect the pest incidence in grain legumes.

It has been reported that chickpea and pigeon pea sown in different months showed differential incidence of insect pests. For example, pigeon pea sown in August had maximum incidence of insect pests than the one sown in September. Similar results have been observed for chickpea, with maximum leaf damage by *H. armigera* and *Spodoptera exigua* Hub. in October sown crop and lowest in January sown crop. Further, stress in plants on account of climate change will result in pest outbreaks as the plant defence system is lowered due to changes in physiological processes.

Table 4: Insect Pests that have become or are likely to become Serious Pests Due to Climate Change

Insect Pest	*Scientific name*	*Crop(s)*
American bollworm	*Helicoverpa armigera* (Hubner)	Cotton, chickpea, pigeon pea,
Beet armyworm	*Spodoptera exigua* (Hub.)	Chickpea in southern India
Spotted pod borer	*Maruca vitrata* (Fabricius)	Pigeon pea, cowpea, lab-lab beans
Sugarcane aphid	*Ceratovacuna lanigera* (Zehnt.)	Sugar cane
Pink stem borer		Maize, sorghum, wheat
Whitefly	*Bemisia tabaci* (Gen.)	Cotton, tobacco
Brown plant-hopper	*Nilaparvata lugens* (Stal)	Rice
Green leaf-hopper	*Nephotettix* spp.	Rice
Serpentine leaf miner	*Liriomyza trifolii* (Burg.)	Cotton, tomato, cucurbits
Wheat aphid	*Macrosiphum miscanthi*	Wheat, barley, oats

Source: Prasad and Bambawale (2010), Fand *et al.* (2012) and Sharma (2013).

Climate Change Effects on Geographic Distribution and Population Dynamics of Insects

Low temperatures are often more important than high temperatures in determining geographical distribution of insect pests. Spatial shifts in distribution of crops under changing climatic conditions will also influence the distribution of insect pests in a geographical region (Parry and Carter 1989). There are several examples of change in the geographic distribution of several insect species as a result of climate change. Populations of the corn earworm [*Heliothis zea* (Boddie)] in the North America might move to higher latitudes/altitudes, leading to greater damage in maize and other crops (EPA 1989). The cotton bollworm/legume pod borers, *H. armigera* and *M. vitrata* will move to temperate regions in northern Europe. New emerging species often spread into completely new ecological settings where most of their natural enemies are missing. An example is the introduction of the B-biotype (*B. tabaci*) in Brazil which was responsible of vectoring of viruses present in native plants only onto cultured tomato crops, leading to new virus diseases in this crop (Fernandes *et al.*, 2008).

Global Warming and Pest Outbreaks

Global warming will result in pest outbreaks (Table 5). Pest outbreaks are more likely to occur with stressed plant's as a result of weakening of plants' defensive system, and thus, increasing the level of susceptibility to insect pests. Global warming will lead to earlier infestation by *H. armigera* in North India (Sharma 2010, 2013), resulting in increased crop loss.

Table 5: Recent Pest Outbreaks in Relation to Climate Change in India

Insect pest	*Host Plant/s*	*Region/ Location*	*Probable Reason/s*	*Impact*	*Reference*
Sugarcane woolly aphid	Sugarcane	Karnataka and Maharashtra during 2002-03	Abnormal weather conditions Insecticide misuse	30 % yield losses Reduced cane recovery	Joshi and Viraktamath (2004), Srikanth (2007)
Nilparvata lugens *Sogatella furcifera*	Rice	North India	Abnormal weather conditions Insecticide misuse	Crop failure on >33,000 ha	IARI (2008)
Phenacoccus solenopsis	Cotton, Vegetables	Punjab, Haryana	Hot and dry weather Insecticide misuse	30-40 % loss	Dhawan *et al.* (2007)
Paracoccus marginatus	Papaya	Tamil Nadu, Karnataka, Maharashtra	Abnormal weather conditions Misuse of insecticides	Significant yield loss	Tanwar *et al.* (2010)

Climate Change Effects on Expression of Resistance to Insect Pests

Host plant resistance to insects is one of the most environmental friendly components of pest management. However, climate change may alter the interactions between the insect pests and their host plants (Sharma 2012b). Resistance to sorghum midge, observed in India, breaks down under high humidity and moderate temperatures in Kenya (Sharma *et al.*, 1999). Cotton bollworm, *Heliothis virescens* (F.) destroyed *Bt* cottons due to high temperatures in Texas, USA (Kaiser 1996). Similarly, *H. armigera* destroyed the cotton crop in the second half of the growing season in Australia because of reduced production of *Bt* toxins in the transgenic crops. Possible causes for the failure of insect control may be inadequate production of the toxin protein, effect of environment on transgene expression, locally resistant insect populations and development of resistance due to inadequate management (Sharma and Ortiz, 2000).

Climate Change Effects on Activity and Abundance of Natural Enemies

Relationships between insect pests and their natural enemies will change as a result of global warming, resulting in both increases and decreases in the status of individual pest species. The interactions between insect pests and their natural enemies need to be studied carefully to devise appropriate methods for using natural enemies in pest management.

Climate Change Effects on Efficacy of Bio-pesticides and Synthetic Insecticides

Natural plant products, entomopathogenic viruses, fungi, bacteria, and nematodes, and synthetic pesticides are highly sensitive to the environment. Increase in temperatures and UV radiation, and a decrease in relative humidity may render many of these control tactics to be less effective, and such an effect will be more pronounced on natural plant products and the bio-pesticides. Therefore, there is a need to develop appropriate strategies for pest management that will be effective under situations of global warming in future. Farmers will need a set of pest control strategies that can produce sustainable yields under climatic change.

Strategies to Mitigate the Effects of Climate Change

Cultural practices, natural enemies, host plant resistance, bio-pesticides, and synthetic pesticides are now being widely used for pest management. However, many of these methods of pest control are highly sensitive to the environment. Therefore, there is a need to develop appropriate strategies for pest management that will be effective under situations of global warming in future.

- ☆ Host plant resistance to insects is one of the most environmental friendly components of pest management. It is important to identify and develop cultivars that are stable in expression of resistance to the target pests under variable climate.
- ☆ Transgenic crops have been developed for controlling some of the most intractable pest problems. There is need to combine host plant resistance from the germplasm with the transgene expression that is stable across environments and locations.
- ☆ Crop diversification is one of the most effective methods of increasing the activity an abundance of natural enemies. There is need to develop crop cultivars that are hospitable to the natural enemies, in addition to identifying cropping systems that will enhance the diversity of natural enemies for pest management.
- ☆ There is a need for a greater understanding of the effect of climate change on the efficacy of synthetic insecticides and their persistence in the environment. Therefore, there is need to develop pesticide formulations and the application schedules that will be least affected by climate change.

☆ Weather based pest forewarning systems are decision support tools that help growers to assess the risk of outbreaks of economically damaging crop pests at present and future climate change scenario. Pest-warning systems are key elements of Integrated Pest Management (IPM).

Conclusion

If ecological knowledge and farmer experience are lacking for these new situations, unexpected outbreaks may become more frequent or more difficult to manage. Taking into account the uncertainty of local and temporal climate change effects in different regions of the world on plant characteristics and crop production, pest and disease epidemiology, and infection probabilities (Chakraborty, 2013), designing appropriate and reliable crop protection strategies will be a real challenge. Hence, the possible effects of climate change on pest evolution, as described earlier, raise several questions concerning the consequences for our current crop protection tools and approaches.

Although climate change is expected to impact Indian agriculture, by influencing the stability of crop yield, information on the real consequences are lacking, and studies in this regard remain in their infancy. Our inability to make confident predictions highlights a need for new research on several fronts. Policies are needed to fill this uncertainty gap by considering a wide range of possible scenarios. Hence, research based on a broader collaborative approach should be made in order to develop anticipatory adaptive strategies resulting in more resilient cropping strategies and stabilized yields. Toward this ultimate aim, enhanced inter-regional, transnational and global networking of researchers and stakeholders, at all levels across the plant health/crop protection spectrum, considering the scale and increasing pressures on crop production and yields, is an effective way to better use the limited resources in order to address this twenty-first century challenge in an ever increasing integrated global context.

References

Awmack, C.S., 2000, Climate change may increase vulnerability of aphids to natural enemies. Ecol. Entomol. 22: 366-368.

Bale, J.S., Masters, G.L. and Hodkinson, I.D., 2002, Herbivory in global climate change research: Directeffect of risising temperature on insect herbivorous. Global climate change Biol. 8: 1-16.

Bezemer, T.M., Jones, T.H. and Knight, K.I., 1999, Long-term effects on elevated carbon dioxide and temperature on peach aphid myzus persicae and its parasitoids *Aphidium matricariae*. Oecologia (Berl) 116: 128-35.

Chakraborty S., Newton, A.C., 2011, Climate change, plant diseases and food security: an overview. Plant Pathol.60: 2-14.

Chakraborty, S., 2013, Migrate or evolve: options for plant pathogens under climate change. Glob Chang Biol. 19: 1985-2000.

Chen, F.G.F., and Parajulee, M.N., 2005, Impact of elevated CO_2 on tritrophic interaction of *Gossypium hirsutum*, Aphid gossypi and Leis axyridis. Env.

Entomol. 34: 37-46.

Coakley, S.M., Scherm, H. and Chakraborty, S., 1999, Climate change and plant disease management.Annual Review of Phytopathology 37: 399-426.

Coviella, C. and Trumble, J.T., 1999, Effects of elevated atmospheric carbon dioxide on insect plant interactions Conserv. Biol. 13: 700-712.

Coviella, C. and Trumble, J.T., 2000, Effect of elevated atmospheric CO_2 on use of foliar application of *Bacillus thuringiensis*. Biocontrol.45: 325-36.

Coviella, C., Margon, D.J.W. and Trumble, J.T., 2000, Interaction of elevated CO_2 and nitrogen fertilization: Effect on the production of *Bacillus thuringiensis* toxins in transgenic plants. Env. Entomol. 29: 281-7.

Dhaliwal, G. S., Jindal, V. and Dhawan, A. K., 2015, Insect pest problems and crop losses: Changing trends. Indian Journal of Ecology. 37: 1-7.

Dhawan, A. K., Singh, K, Saini, S., Mohindru, B., Kaur, A., Singh, G. and Singh, S., 2007, Incidence and damage potential of mealybug, *Phenacoccus solenopsis* Tinsley, on cotton in Punjab. Indian Journal of Ecology. 34 : 110-116.

Dukes, J.S. and Mooney, H.A., 1999, Does global change increase the success of biological invaders? Trends. Ecol. Evol. 14: 135-9.

Epstein, P.R., Diaz, H.F., Flias, S., 1998, Biological and physical sign of climate change: Focus Mosquito-borne diseases. Bull. Am. Meteorol. Soc. 79: 409-17.

Erasmus, B.F.N., Van Jarsveld, A.S. and Chown, S., 2002, Vulnerability of South Africa animal taxa on climate change. Global Change Boil.6: 679-693.

Fand, B. B., Kamble, A. L. and Kumar, M., 2012, Will climate change pose serious threat to crop pest management: A critical review? International Journal of Scientific Research Publications 2 : 2250-3153.

Fernandes, F.R., deAlbuquerque, L.C., Giordano, L.D. and Betal, 2008, Diversity and prevalence of Brazilian bipartite bergomovirus species associated to tomoatoes. Virus Gene 36: 251-258.

Fleming, R.A. and Volney, W.J.A., 1995, Effect of climate change on insect defoliator population process in Canada boreal forest- some plausible scenarios. Water Air Soil Pollution 82: 445-54.

Fletcher, J, Luster, D., Bostock Retal, 2010, Emerging infectious diseases. In: Scheld WM (ed) Emerging Infections. ASM Press, Washington DC, pp. 337-366.

Guo, H., Sun, Y. and Li, Y., 2013, Pea aphid promotes amino acid metabolism both I Medicago truncatula and bacteriocytes to favor aphid population growth under elevated CO_2. Glob. Chang Biol. 19: 3210-3223.

Hance, T., Van Baaren, J., Vernon, P. and Bolvin, G., 2007, Impact of streme temperature on parasitoids in a climate change perspective. Ann. Rev. Entomol. 52: 107-26.

Hunter, M.D., 2001, Effects of elevated atmospheric carbon dioxide on insect-plant interactions. Agric. For. Entomol. 3: 153-159.

Indian Agricultural Research Institute News, 2008, Brown plant hopper outbreak in rice. IARI News 24: 1-2.

Joshi, S. and Viraktamath, C. A., 2004, The sugarcane woolly aphid, *Ceratovacuna lanigera* Zehntner (Hemiptera: Aphididae): its biology, pest status and control. Current Science 87: 307-316.

Kaiser, J., 1996, Pests overwhelm Bt cotton crop. Nature., 273: 423.

Lastuvka, Z., 2009, Climate change and its possible influence on the occurrence and importance of insect pests. Plant Protect. Sci. 45: S53-S62.

Liu, D. and J. T., Trumble, 2007, "Comparative fitness of invasive and native populations of the potato psyllid (*Bactericera cockerelli*)." Entomologia Experimentalis et Applicata. 123(1): 35-42.

Moore, B.A., Allard, G.B., 2008, Climate change impacts on forest health. Forest Health and Biosecurity Working Papers, Forestry Department Food and Agriculture Organization of the United Nations. Working Paper FBS/9E, FAO, Rome, Italy. pp 35.

Osbrink, W.L.A., Trumple, J.T. and Wagner, R.E., 1987, Host suitability of *Phaseolus lunata* for Trichoplusiani (Lepidoptora: Noctuidae) in controlled carbon dioxide atmosphere. Envi. Entomol. 16: 639-44.

Parmesan, C., 2006, Ecological and evolutionary response to recent climate change. Ann. Rev. Ecol. Syst. 37: 637-63.

Parmesan, C., 2007, Influences of species, latitudes and methodologies on estimates of phonological response to global warming. Global Change Biol. 13: 1860-1872.

Prasad, Y.G. and Bambawale, O. M., 2010, Effects of Climate Change on Natural Control of Insect Pests. Indian Journal of Dryland Agriculture Research and Development 25: 1-12.

Regnier, J., 1983, An oviposition model for the spruce budworm, choristoneura fumiferana (Lepidoptera: Tortricidae). Can. Entomol. 115: 1371-82.

Roth, S.K. and Lindroth, R.L., 1995, Elevated atmospheric CO_2: Effect on photochemistry, insect performance and insect parasitoid interactions. Global Change Biol. 1: 173-82.

Sharma, H. C., Mukuru, S. Z, Manyasa, E. and Were, J., 1999, Breakdown of resistance to sorghum midge, *Stenodiplosis sorghicola*. Euphytica 109: 131-140.

Srikanth, J. 2007, World and Indian scenario of sugarcane woolly aphid. In: Woolly Aphid Management in Sugarcane (Mukunthan N, Srikanth J, Singaravelu B, Rajula Shanthy T, Thiagarajan R and Puthira Prathap D, eds.). Extension Publication, Sugarcane Breeding Institute, Coimbatore, Tamil Nadu, India. Pp1-12.

Sharma, H. C., 2012, Climate change effects on activity and abundance of insects: Implications for Crop Protection and Food Security. In: Kang MS, Banga SS (Ed) Combating Climate Change: An Agricultural Perspective. Taylor and Francis, Boca Raton, Florida, USA.

Sharma, H. C., Ortiz, R., 2000, Transgenics, pest management and the environment. Current Science 79: 421-437.

Stacey, D. A. and Fellows, M.D.E., 2002, Influence of temperature on pea aphid A cythosiphan pisum (hemiptera: aphidae) resistence to natural enemy attack. Bull. Entomol. Res. 92: 351-7.

Tanwar, R. K., Jeyakumar, P. and Vennila, S., 2010, Papaya mealybug and its management strategies. Technical Bulletin 22. National Centre for Integrated Pest Management, New Delhi-110 012, India.

Thomas, C.D., Cameran, A. and Green R.E., 2004, Extinction risk from climate change. Nature. 427: 145-8.

Thomson, L.J., Macfadyen, S. and Hoffmann. A.A., 2010, Predicting the effects of climate change on natural enemies of agricultural pests. Biol. Control.52: 296-306.

Woiwod, I.P. and Harrington, R., 1994, "Flying in the face of change: the Rothamsted Insect Survey. Long-Term Experiments in Agricultural and Ecological Sciences" (eds R.A.Leigh & A.E.Johnston). CAB International, London, UK. pp. 321-337.

Yamamura, K. and Kiritani, K., 1998, A simple method to estimate the potential increase in the number of generations under global warming in temperate zones. Appl. Entomology Zool. 33: 289298.

Zhou, X.L., Harrington, R. and Woiwod, I.P., 1995, Effects of temperature on aphid phenology. Global Chang Biol 1: 303–313.

4

Impact of Climate Change on Insect and Mite Pests in Horticultural Crops

V. Sridhar

Introduction

In agriculture and horticulture ecosystems, weather determines the phenology and dynamics of species. Weather affects crop yield and quality as well as the dynamics of pests and their regulation by natural enemies (De Bach, 1964). The term climate change embraces a range of natural and anthropogenic environmental changes. According to Intergovernmental Panel on Climate Change, it is defined as "Change in climate over time, either due to natural variability or as a result of human activity". The most of the warming observed over the last 50 years is attributable to human activities. The global mean surface temperature is predicted to increase by 1.4 to 5.8°C from 1990 to 2100. If temperatures rise by about 2°C over the next 100 years, negative effects of global warming would begin to extend to most regions of the world (IPCC, 2001). Such changes in climate and weather could profoundly affect the population dynamics and status of insect pests of crops. These may arise not only as a result of direct effects on the distribution and abundance of pest populations but also indirect effects on the pests' host plants, competitors and natural enemies (Porter, *et al.*, 1991). Some pests which are already present but only occur in small areas, or at low densities may be able to exploit the changing conditions by spreading more widely and reaching damaging population densities (Bale *et al.*, 2002).

Global warming and climate changes will result in following phenomena directly influencing the insects or mites or indirectly through the host plants.

- ☆ Extension of geographical range of insect pests.
- ☆ Impact on arthropod diversity and extinction of species.
- ☆ Change on the biology and population dynamics of insect pests.
- ☆ Change on species interactions.
- ☆ Increased over-wintering and rapid population growth.
- ☆ Evolutionary changes and emerging pest problems.
- ☆ Effects on existing pest management strategies.

Extension of Geographical Range of Pests

Environmental factors influenced by global climate change determine the distribution ranges of organisms. It plays a major role in defining the distribution limits of a species. Due to climate change, it is predicted that the distribution of most insect species will shift towards the poles and to higher elevations with predicted temperature increase and temperate regions will bear the main burden of these shifts. With changes in climate, these limits are shifting as species expand into higher latitudes and altitudes and disappear from areas that have become climatically unsuitable (Parmesan, 2006 and Menendez, 2007). Climate change will alter the ranges and abundances of insects and therefore have profound impacts on crop cultivation by the movement of existing crop pests into new areas and potentially, by rising currently disregarded insect species to pest status.

Zalucki and Furlong (2008) predicted the regional difference in the occurrence of *Plutella xylostella* in Australia by CLIMEX models suggesting by 2090 there will be a substantial decrease in the areas suitable for year-round growth and development of *P. xylostella* across Australia mainly due to temperatures increases. However, they observed the regional differences and in general, areas in the coastal fringe become more suitable, while areas inland and in Northern Australia will become less suitable for this species.

Sridhar *et al.*(2014)made a study to understand the influence of climate change on future potential distribution of Oriental fruit fly (OFF), *Bactrocera dorsalis* (Hendel) for different time-frames (2030, 2050, 2070 and 2090) under the A1B climate change scenario (CSIRO-Mk3.0, a global climate model) using CLIMEX software. They have reported that an overall gradual decrease in terms of area suitability for OFF in India by 2090 due to increase in various stress factors to the pest. In temperate regions of the north and north-eastern parts, incidence of the pest may increase due to rise in temperature coupled with decrease in cold stress.

Climate Change and Arthropod Diversity and Species Extinction

Large-scale changes in rainfall caused by global warming will also have a major bearing on the abundance and diversity of arthropods. Extreme climatic events, such as drought, are likely to decrease multi-trophic diversity and change the composition of arthropod communities, which, in turn, may affect other associated taxa. Arthropods (insects, spiders, and mites) are the most abundant and diverse group of organisms (Kannan and James 2009; Gregory *et al.*, 2009). The main effects of climate change and pollution on arthropod communities will be a decreased abundance of decomposers and natural enemies and increased herbivory, which may have negative consequences for the structure and services of entire ecosystems. Climate change will become a major factor for the extinction of arthropod species (Thomas *et al.*, 2004; Butchart *et al.*, 2005). Mountain species and those restricted to high latitudes are most likely to become extinct as a result of climate change. The species adapted to cold conditions will be forced to move uphill to higher latitudes as a result of global warming. Some of the species may eventually run out of habitable areas and inevitably become extinct. Species decline is difficult to detect and takes place across several hundreds of years.

Four butterfly species have disappeared from lower latitudes during the past 25 years in the United Kingdom, with climate change being responsible for atleast half of the population extinctions (Franco *et al.*, 2006). In Spain, lower elevation limits of 16 mountain butterflies have increased in altitude by around 212 m as a result of global warming (Wilson *et al.*, 2005). Speciation takes between 100 and 1,000,000 years, resulting in 10-10,000 new species per annum. Nearly 99.9 % of all species that ever existed have become extinct.

Change in the Biology and Population Dynamics of Insect Pests

Higher temperatures below the species' upper threshold limit will result in rapid development and increases in pest populations as the time to reproductive maturity will be reduced considerably. In addition to the direct effects of temperature changes on development rates, improvement in food quality because of abiotic stresses may result in increases in the growth rates of some insect species (White 1984) and decrease in growth rates in others (Maffei *et al.*, 2007). The calculated hypothetical number of generations per year of *Hypothenemus hampei* in Eastern Africais predicted to increase in all *C. arabica*-producing areas from five to ten. These outcomes will have serious implications for *C. arabica* production and livelihoods (Jaramillo *et al.*, 2011). Sridhar and Reddy (2013) reviewed the use of degree day and plant phenology as a reliable to for prediction of insect pest activity under climate change conditions.

Joutei *et al.*, (2000) studied influence of elevated CO_2 on biology of two-spotted spider mites (*Tetranychus urticae*) and reported that mites raised on common beans grown at 600 and 700 ppm CO_2 and significant decrease in the number of the offspring in the first and second generations (34% and 49%) respectively was observed compared to ambient CO_2. Similar experiment was conducted, where *T. urticae* were raised on clover (*Trifolium repens*) grown at different CO_2 levels (395-748 ppm). The results showed a quite opposite effect: under elevated CO_2 spider mite reproduction increased significantly compared to lower CO_2 (Heagle *et al.*, 2002).

Tomato plants grown under eCO_2 (550 ppm) were found to be significantly less susceptible to *L. trifolii* than that of ambient grown plants in both choice (P<0.001: F=64.5) and no choice (P<0.001: F=152.87) assays. Tomato fruit borers, *Helicoverpa armigera* and *Spodoptera litura* prefer eCO_2 grown tomato plants over ambient grown tomato plants; gravid female showed significantly more egg laying (Sridhar *et al.*, unpublished).

Studies conducted by various workers on the impact of air temperature on number of generations of horticultural crop pests is summarised in the Table 1.

Table 1: Impact of Air Temperature on Number of Generations of Insect Pests

Insect	*Crop*	*Observation*	*Reference*
Plutella xylostella	Brassicas	2° increase in mean temperature. Forecast the appearance of two new generations per year	Mori moto *et al* (1998)
P. xylostella, Helicoverpa armigera, Trichoplusiani	Brassicas & tomato	Increase in *H. armigera* and *Trichoplusiani* and decline in *P. xylostella*	Kiritani, 2006
Grape Berry Moth, *Paralobesia viteana*	Grape	Increase in mean surface temperature more than 2°C had dramatic effect on insect voltinism	Tobin *et al.*, 2008
Bark beetles, *Ipstypographus*	Spruce	Accelerated developmental rates with early completion of life cycle and additional generations within a season	Lange *et al*, 2006 Jonssonv *et al.*, 2009
Peach fruit fly	Peach	More number of generations in future climate during 2050 and 2100	Khalil *et al.*, 2010
P. xylostella	Brassicas	More number of generations in tropical region and was nearly twice the number than in sub-tropical region of Brazil	Marchioro and Foerster 2011
Potato Tuber worm, *Phthorimaeaoperculella*	Potato	Higher number of generations per year during 2050 (9-11) and 2100 (10-12).	Abolmaaty *et al.*, 2011
P. xylostella	Cole crops	Appearance of two new generations per year	Kocsis and Hufnegal, 2011
Cydia pomonella	Apple	The risk of an additional third generation increase by 2-100 %. Shift in the phenology and voltinism were predicted	Sibylle Stoecli *et. al.*, 2012

Effect of Climate Change on Species Interactions

Climate-driven range expansion has created new interactions with potentially devastating consequences for crop production and food security. Changes in phenological patterns and distribution of individual species will result in altered species interactions within communities. The magnitude of response will differ between species, and there is considerable potential for the disruption of existing interactions among species (Visser *et al.*, 2004). Interactions that involve two or more trophic groups, such as plant–herbivore, plant– pollinator, and insect–host–natural enemy interaction are likely to undergo a major change (Harrington *et al.*, 1999). Some plant species may be unable to cope with climate change, resulting in the extinction of species that are specific to particular host plants (Thomas *et al.*, 2004).

Distributional changes are also expected to bring about changes in species interactions because some species are likely to expand their distribution to different areas, for example, the pine processionary moth, *Thaumetopoea pityocampa* has moved to higher elevations during the last 20 years as a result of increasing temperature and has encountered a new host, Scots pine (*Pinus sylvestris* [L.] var. Nevadians *H. Christ.*). In warmer years, the defoliation by this insect has increased from 5%–25% above 1700 m (Hodar and Zamora, 2004). An increase of 3°C in mean daily temperature would cause the carrot fly, *Delia radicum* (L.), to become active a month earlier than at present (Collier *et al.*, 1991) and temperature increases of 5°C–10°C would result in the completion of four generations each year compared with two to three generations presently (Biron *et al.*, 2000), necessitating the adoption of new pest control strategies. Global warming will lead to earlier infestation by insect pests, such as *Helicoverpa zea* (Boddie) in North America (EPA 1989) and *Helicoverpa armigera* (Hub.) in north India (Sharma, 2010), resulting in increased crop losses. Rising temperatures are likely to result in the availability of new niches for insect pests. The emergence of some insect pests from diapause now occurs much earlier than that of their natural enemies, resulting in a mismatch between the interacting species.

Increased Over-Wintering and Rapid Population Growth

As poikilothermic animals, insects have limited ability to regulate their body temperature and whilst this lack of internal thermal homeostasis might be viewed as disadvantageous. Likely impacts of a changing climate on the low temperature biology of insects is clearly a complex subject but there are three factors that are particularly important to consider: the cold tolerance of species under prevailing climatic conditions, the occurrence of diapause, and the direct and indirect effects of higher temperatures on one or both of these overwintering processes (Bale and Hayward, 2009).Overwintering of insect pests will increase as a result of climate change, producing larger spring populations as a base for a build-up in numbers in the following season. These may be vulnerable to parasitoids and predators if the latter also overwinter more readily. Temperature has a strong influence on the viability and incubation period of *H. armigera* eggs, which can be predicted on the basis of degree-days required for egg hatching (Dhillon and Sharma, 2007).

One of the problems of trying to place the cold hardiness of insects in general within the alternative strategies of freeze tolerance and avoidance is that these strategies were derived from studies on cold hardy insects living in cold and very cold climates, whereas the majority of species live in much less severe climates, and biodiversity is greatest in the tropics. A range of studies have shown that whilst most insects have considerable potential to 'supercool', a low super cooling points (SCPs) is not necessarily indicative of cold hardiness. For example, there are many species with winter SCPs of –25°C but when held at higher sub-zero temperatures, mortality increases after exposures of 1-3 months [*e.g.* the beech weevil *Rhynchaenus fagi* L. (Bale, 1991)] or only a few minutes *e.g.* the aphid *Myzus persicae* (Sulz.) (Bale *et al.*, 1988; Clough *et al.*, 1990). With species of tropical and Mediterranean origin, high mortality can occur whenever the insects experience temperatures below the developmental threshold, *i.e.* between 10°C and 0°C (Hatherly *et al.*, 2005); hence, overwintering of such species in cool temperate climates is dependent on a high level of thermal protection. With respect to the effects of cold exposure, Bale described species such as *R. fagi*, *M. persicae* and tropical insects as 'chill tolerant' 'chill susceptible' and 'opportunistic survivors', respectively. In reality, the terminology is not important – the key point is that most insects are neither freeze tolerant nor freeze avoiding (because they are killed by the effects of cold, not freezing), and this diversity of thermal tolerance needs to be recognised in any generalised consideration of the impacts of climate warming on insect winter survival (Bale, 1993; Bale, 1996).

Evolutionary Changes and Emerging Pest Problems

Global warming may change the flowering times of plants, leading to ecological consequences, such as the introduction of new insect pests and the attainment of pest status by minor insect pests (Willis *et al.*, 2008). Climate change will alter selection pressures within populations because most populations are adapted to their local environments. Traits that confer fitness in an environment might not be as successful under new climatic conditions, and therefore it is imperative that evolutionary responses take place. The potential of evolutionary changes is quite high among insects as they grow fast, have short generation times, and have high reproductive rates. Insects will adjust rapidly to new environmental conditions because of faster generation turnover than other animals. Therefore, it is not surprising that most empirical evidence of evolutionary changes has come from insect species (Parmesan, 2006). Adaptation to climate change has also been observed in species with high frequencies of pre-existing high-temperature-tolerant genotypes.

Chromosomal inversion polymorphism in Drosophila species changes with latitude and is related to climatic factors. Under warm climatic conditions, the frequency of high-temperature-adapted genotypes increases considerably within existing populations (Levitan, 2003; Balanya *et al.*, 2006).

Effects on Existing Pest Management Strategies

As a result of global warming, insect pests confined to tropical and subtropical regions may move to temperate regions along with a shift in the areas of production of their host plants and it affects the existing pest management strategies due to change in the expression of host plant resistance to insects, Affect the efficacy of insect-resistant transgenic crops, Affect the activity and abundance of natural enemies, Reduce the effectiveness of bio-pesticides and synthetic pesticides for pest management.

Elevated CO_2 and Expression of Resistance to Insect Pests

CO_2 impacts on insects are thought to be indirect. Impact on insect damage will result from changes in nutritional quality and secondary metabolites of the host plants. Increased levels of CO_2 will enhance plant growth, but may also increase the damage caused by some phytophagous insects. In the enriched CO_2 atmosphere expected in the 21st century, many species of herbivorous insects will confront less nutritious host plants that will induce both lengthened larval developmental times and greater mortality (Coviella and Trumble, 1999). The effects of climate change on the magnitude of herbivory and direction of response will not only be species-specific, but also specific to each insect–plant system. Bark beetles, wood borers, and sap sucking insects benefit from severe drought (Huberty and Denno, 2004;

Koricheva *et al.*, 1998), while *Spodoptera exigua* (Hub.) exhibited a reduced ability to feed on drought-stressed tomato leaf tissue, which contained higher levels of defence compounds as a result of the abiotic stress (English-Loeb *et al.*, 1997).

Increased CO_2 may also cause a slight decrease in nitrogen-based defences (*e.g.* alkaloids) and a slight increase in carbon-based defences (*e.g.* tannins). Lower foliar nitrogen due to CO_2 causes an increase in food consumption by herbivores. Soybeans grown in elevated CO_2 suffered 57 % more damage from herbivores (primarily Japanese beetle, potato leafhopper, western corn rootworm and Mexican bean beetle) than those grown in ambient CO_2. Increase in amounts of simple sugars and down-regulation of gene expression for a protease-specific deterrent to coleopteran herbivores may have resulted in greater insect feeding (Zavala *et al.*, 2008).

Transgenic Crops for Pest Management

Transgenic plants expressing the *Bacillus thuringiensis* (Bt) (Berliner) insecticidal protein showed a reduction in the level of toxin protein during periods of high temperature, elevated CO_2 levels, or drought, leading to decreased resistance to insect pests (Chen *et al.*, 2005; Dong and Li, 2007). Cotton bollworm, *Heliothis virescens* (F.), destroyed Bt cottons due to high temperatures in Texas, USA (Kaiser, 1996). Similarly, *H. armigera* and *H. punctigera* damaged *Bt*-cotton in the second half of the growing season in Australia because of reduced production of *Bt* toxins in the transgenic crops (Hilder and Boulter, 1999). Cry1Ac levels decrease with plant age, resulting in greater susceptibility of the crop to bollworms during the later stages of crop growth (Adamczyk *et al.*, 2001; Greenplate *et al.*, 2000; Kranthi *et al.*, 2005). Possible causes for the failure of insect control may be due to inadequate production of the toxin protein, the effect of environment on transgene expression, locally resistant insect populations, and development of resistance due to inadequate management (Sharma and Ortiz, 2000).

Activity and Abundance of Natural Enemies

Increases in atmospheric CO_2, low precipitation and increases in temperature will alter plant phenology, influencing herbivore growth and abundance, and indirectly affecting the abundance of prey and insect hosts for natural enemies (Thomson *et al.*, 2010). Relationships between insect pests and their natural enemies will change as a result of climate change, resulting in both increases and decreases in the status of individual pest species.

Changes in temperature will also alter the timing of diurnal activity patterns of different groups of insects (Young, 1982), and changes in interspecific interactions could also alter the effectiveness of natural enemies for pest management (Hill and Dymock, 1989). The fitness of natural enemies will decline as the quality of their herbivore hosts decreases (Wang *et al.*, 2007) as has been shown for several groups of predators including spiders (Hvam and Toft, 2005; Toft, 1995), predatory bugs (Butler and O'Neil, 2007) and carabid beetles (Bilde and Toft, 1999). However, a decrease in prey size will not necessarily always lead to a reduction in the success of predators. The number of prey consumed by predators might increase and lead to improved pest control (Chen *et al.*, 2005; Coll and Hughes, 2008). The pentatomid bug, oechalia schellenbergii Guerin Meneville, exhibited increased predation of cotton bollworm, *H. armigera*, feeding on peas under elevated CO_2 because the pea plants had reduced nitrogen content when grown under high CO_2, which influenced the size of the cotton bollworm larvae (Coll and Hughes, 2008). At higher temperatures, aphids were less responsive to the aphid alarm pheromone they release when under attack by insect predators and parasitoids, resulting in the potential for greater predation (Awmack *et al.*, 1997).

Efficacy of Bio-pesticides and Synthetic Insecticides vs Temperature

Temperature is a major factor affecting insecticide toxicity (DeVries and Georghiou, 1979), and, thus, efficacy (Johnson, 1990; Scott, 1995). The effects of temperature on efficacy can be either positive or negative. The response relationship between temperature and efficacy has been found to vary depending on the mode of action of an insecticide, target species, method of application, and quantity of insecticide ingested or contacted (Johnson, 1990). Increased temperature will increase the activity of some of the insecticides. Diflubenzuron (an insect growth regulator (IGR)) caused rapid mortality at higher temperatures and was more efficient at 35°C (Amarasekare and Edelson, 2004). This was probably because this IGR is only effective when the insect moults (Ware, 2000), and the insect growth rate and moulting rate increase at higher temperatures (Lactin and Johnson, 1995).

Increase in temperature and UV radiation, and a decrease in relative humidity, may render many of the pest control tactics to be less effective, and such an effect will be more pronounced on natural plant products and the bio pesticides. Several studies have reported a significant decrease in biological activity of entomopathogens, *viz.* NPV, GV, *Beauveria* and *Bt* (up to 90%) within a few days (Broome *et al.*, 1974; David *et al.*, 1968; Jones and McKinley, 1986). Another effect of increased temperature and UV radiation may be to slow down the activity even without the loss of activity due to UV radiation; as a result, more time may be required to achieve insect mortality. Larvae continue to feed and damage crops until shortly before death (Chen and McCarl, 2001).

Conclusions

Current estimates of climate change indicate an increase in global mean annual temperature and CO_2 concentration and a decrease in annual precipitation, which might result in severe adverse effects on crop cultivation, resulting in changes in crop diversity and the availability of food. Global warming and climate change will inflict major changes in arthropod diversity, geographical distribution of insect pests, insect population dynamics, herbivore-plant interactions, activity and abundance of natural enemies, effectiveness of host plant resistance, transgenic plants, biopesticides, and synthetic chemicals for pest management and efficacy of crop protection technologies in horticulture ecosystem. As a result, economic relationships between the costs and benefits of pest control measures are expected to change. Therefore, there is a need to generate information on the likely effects of climate change on insect pests to develop robust technologies that will be effective and economical for the pest management under global warming and climate change.

References

Abolmaaty S.M., Khalil A.A. and Amna M.H. 2011. Using degree – day unit accumulation to predict potato tuber worm incidence under climate change conditions in Egypt. Nature and Science. 9(4): 156-160.

Adamczyk Jr., J. J., Adams, L.C and Hardee, D.D. 2001. Field efficacy and seasonal expression profiles for terminal leaves of single and double *Bacillus thuringiensis* toxin cotton genotypes. J. Econ. Entomol., 94: 1589-1593.

Amarasekare, K.G. and Edelson, J.V., 2004. Effect of temperature on efficacy of insecticides to differential grasshopper (Orthoptera: Acrididae). J. Econ. Entomol., 97: 1595-1602.

Awmack, C.S., C.M. Woodcock and Harrington., R. 1997.Climate change may increase vulnerability of aphids to natural enemies. Ecologi. Entomol., 22: 366-368.

Balanya, J., Oller, J. M., Huey, R.B., Gilchrist, G.W and Serra. L. 2006. Global genetic change tracks global climate warming in *Drosophila subobscura*, Science, 313: 1773-1775.

Bale J. S. and Hayward S. A. L., 209. Insect overwintering in a changing climate. *The Journal of Experimental Biology*, 213: 980-994.

Bale, J. S. 1993. Classes of insect cold hardiness. *Func. Ecol.*, 7: 751-753.

Bale, J. S. 1996. Insect cold hardiness: a matter of life and death. *Eur. J. Entomol.*, 93: 369-382.

Bale, J. S., Harrington, R. and Clough, M. S. 1988. Low temperature mortality of the peach potato aphid *Myzus persicae*. *Ecol. Entomol.*, 13: 121-129.

Bale, J., G. Masters, I. Hodkinson, C. Awmack, T.M. Jnbezemer, V.K. Brown, J. Butterfield, A. Buse, J.C. Coulson, J. Farrar, J.G. Good, R. Harrington, S. Hartley, T.H. Jones, L. Lindroth, M. Press, I. Mrnioudis, A. Watt and A. Whittaker, 2002. Herbivory in global climate change research: direct effects of rising temperature on insect herbivores. *J. Global Change Biol.*, 8: 1-16.

Broome, J. R., Sikorowski, P. P and Nee, W. W., 1974. Effect of sunlight on the activity of nuclear polyhedrosis virus from *Malacoso madisstria*. *J. Econ. Entomol.* 67: 135-136.

Butchart, S. H. M., Stattersfield, A. J., Bennun, L. A., Akcakaya, H. R., Baillie, J. E. M., Stuart, S.N., Hilton-Taylor, C., and Mace, G. M. 2005. Using red list indices to measure progress towards the 2010 target and beyond. Philosophical Transactions of the Royal Society, London 1454: 255-268.

Chen, C. and McCarl, B., 2001. Pesticide usage as influenced by climate: a statistical investigation. *Clim. Change*, 50: 475-487.

Chen, D.H., Ye, G.Y., Yang, C.Q., Chen, Y. and Wu, Y.K., 2005. The effect of high temperature on the insecticidal properties of *Bt* Cotton. *Environ. Exp. Bot.*, 53: 333–342.

Chen, F.J., Wu, G., Ge, F., Parajulee, M.N. and Shrestha, R.B., 2005a. Effects of elevated CO_2 and transgenic *Bt* cotton on plant chemistry, performance, and feeding of an insect herbivore, the cotton bollworm. *Entomol. Exp. Appl.*, 115: 341-350.

Clough, M. S., Bale, J. S. and Harrington, R. 1990. Differential cold hardiness in adults and nymphs of the peach-potato aphid *Myzus persicae*. *Ann. Appl. Biol.*, 116: 1-9.

Coll, M. and Hughes, L., 2008. Effects of elevated CO_2 on an insect omnivore: a test for nutritional effects mediated by host plants and prey. *Agric. Ecosyst. Environ.*123: 271-279.

Coviella, C.E. and Trumble, J.T., 1999. Effects of elevated atmospheric carbon dioxide on insect-plant interactions. *Conserv. Biol.*, 13: 700-712.

David, W. A. L., Gardiner, B.O.C. and Wooner, M., 1968. The effects of sunlight on a purified granulosis virus of *Pieris brassicae* applied to cabbage leaves. *J. Invertebr. Pathol.*, 11: 496-501.

De Bach, P. 1964. Biological Control of Insect Pests and Weeds, p 350. New York.

DeVries, D. H. and Georghiou, G. P., 1979. Influence of temperature on toxicity of insecticides to susceptible and resistant house flies. *J. Econ. Entomol.*72: 48-50.

Dhillon, M. K., and H. C. Sharma. 2007. Effect of storage temperature and duration on viability of eggs of *Helicoverpa armigera* (Lepidoptera: Noctuidae). *Bull. Entomol. Res.*, 97: 55-59.

Dong, H. Z. and Li, W.J., 2007. Variability of endotoxin expression in *Bt* transgenic cotton. *J. Agron. Crop Sci*, . 193: 21-29.

English-Loeb, G., Stout, M. J. and Duffey, S. S., 1997. Drought stress in tomatoes: changes in plant chemistry and potential nonlinear consequences for insect herbivores. *Oikos*, 79: 456-468.

Franco, A. M. A., Hill, J. K., Kitschke, C., Collingham, Y., Roy, D. B., Fox, R., Huntley, B. and Thomas, C. D. 2006. Impacts of climate warming and habitat loss on extinctions at species' low-latitude range boundaries. Glob. Change Biol., 12: 1545-1553.

Greenplate, J. T., Penn, S. R., Shappley, Z., Oppenhuizen, M., Mann, J. and Reich, B., 2000. Bollgard II efficacy: quantification of total lepidopteran activity in a 2-gene product. In: Dugger, P., Richter, D. (Eds.), Proceedings, Beltwide Cotton Conference National Cotton Council of America, Memphis, TN, USA, pp. 1041–1043.

Gregory, P. J., Johnson, S. N., Newton, A. C and Ingram, J. S. I. 2009. Integrating pests and pathogens into the climate change/food security debate. J. Experi. Botany, 60: 2827-2838.

Harrington, R., Woiwod, I and Sparks, T. 1999. Climate change and trophic interactions. *Trends Ecol. Evol.*, 14: 146-150.

Hatherly, I. S., Hart, A. J., Tullett, A. G. T. and Bale, J. S. 2005. Use of thermal data as a screen for the establishment potential of non-native biocontrol agents in the UK. *Biocontrol*, 50: 687-698.

Heagle, A.A., Burns, J. C., Fisher, D. S. and Miller, J. E. 2002. Effects of carbon dioxide enrichment on leaf chemistry and reproduction by two-spotted spider mites (Acari: Tetranychidae) on white clover. Envl. Entomol., 31: 594-601.

Hilder, V. A and Boulter, D., 1999. Genetic engineering of crop plants for insect resistance-a critical review. *Crop Protect.*, 18: 177-191.

Hill, M.G., Dymock, J.J., 1989. Impact of Climate Change: Agricultural/Horticultural Systems. DSIR Entomology Division Submission to the New Zealand Climate Change Program, Department of Scientific and Industrial Research. Auckland, New Zealand, p. 16.

Hodar, J. A., and R. Zamora. 2004. Herbivory and climatic warming: a Mediterranean outbreaking caterpillar attacks a relict boreal pine species. *Biodivers. Conserv.* 13: 493-500.

Huberty, A. F and Denno, R. F., 2004. Plant water stress and its consequences for herbivorous insects: a new synthesis. *Ecology*, 85: 1383-1398.

IPCC, 2001. Climate change 2001: the scientific basis, Contribution of working group I to the third assessment report of the intergovernmental panel on climate change (IPCC). Cambridge University Press, Cambridge http ://www. grida. no/climate/ipcc_tar/.

Jaramillo, J., Muchugu, E., Vega, F. E., Davis, A., Borgemeister, C and Chabi-Olaye, A., 2011. Some Like It Hot: The Influence and Implications of Climate Change on Coffee Berry Borer (*Hypothenemus hampei*) and Coffee Production in East Africa. PLoS ONE 6(9): e24528. https://doi.org/10.1371/journal.pone.0024528.

Johnson, D. L. 1990. Influence of temperature on toxicity of two pyrethroids to grasshoppers (Orthoptera: Acrididae). *J. Econ. Entomol.*83: 366-373.

Jones, K. A and McKinley, D. J. 1986. UV inactivation of *Spodoptera littoralis* nuclear polyhedrosis virus in Egypt: assessment and protection. In: Samson, R.A. Vlak, J.M., Peters, D. (Eds.), Fundamental and Applied Aspects of Invertebrate Pathology. Proceedings, IV International Colloqium on Invertebrate Pathology, Wageningen, The Netherlands, p. 155.

Joutei, A. B., Roy, J., van Impe, G. and Lebrun, P. 2000. Effect of elevated CO_2 on the demography of a leaf-sucking mite feeding on bean. *Oecologia*, 123: 75-81.

Kaiser, J., 1996. Pests overwhelm *Bt* cotton crop. Nature 273, 423.

Kannan, R. and James, D.A. 2009. Effects of climate change on global diversity: a review of key literature. Tropical Ecology, 50: 31-39.

Khalil, A. A., Shaker, M. Abolmaaty, Mousad, K., Hassancin, M., ostafa M. El-Mtewally and S. Ameh A. Moustafa. 2010. Degree-days units and expected generation numbers of peach fruit by *Bactrocera zonata* (Saunders) (Diptera: Tephritidae) under climate change in Egypt. Egypt. Acad. J. biolog. Sci., 3(1): 11-19.

Kiritani, K. 2006 Predicting impacts of global warming on population dynamics and dynamics and distribution of arthropods in Japan. Popul. Ecol.48: 5-12.

Kocsis, S. M. and Hufnagel, L. 2011. Impacts of Climate Change on Lepidoptera Species and Communities. Appl. ecology and environmental research, 9 (1): 43-72.

Koricheva, J., Larsson, S. and Haukioja, E., 1998. Insect performance on experimentally stressed woody plants: a meta-analysis. Ann. Rev. Entomol., 43: 195-216.

Kranthi, K. R., Naidu, S., Dhawad, C.S., Tatwawadi, A., Mate, K., and Patil, E., 2005. Temporal and intraplant variability of Cry1Ac expression in Bt-cotton and its influence on the survival of the cotton bollworm, *Helicoverpa armigera* (Hubner) (Noctuidae: Lepidoptera). Curr. Sci., 89: 291-298.

Lactin, D. J. and Johnson, D. L., 1995. Temperature dependent feeding rates of *Melanoplus sanguinipes* nymphs (Orthoptera: Acrididae) in laboratory trials. Environ. Entomol., 24: 1291-1296.

Lange, H., B. Okland and P. Krokene, 2006. Thresholds in the life cycle of the spruce bark beetle under climate change. Int. J. for Complex Systems, 1648L 1-10.

Levitan, M. 2003. Climatic factors and increased frequencies of 'southern' chromosom forms in natural populations of Drosophila robusta. Evol. Ecol. Res.

Maffei, M. E., A. Mithofer, and W. Boland. 2007. Insects feeding on plants: Rapid signals and responses proceeding induction of phytochemical release. Phytochemistry 68: 2946-2959.

Marchioro, C.A. and Foerster L.A. 2011. Ecology, Behavior and Bionomics Development and survival of the Diamondback Moth, *Plutella xylostella* (L.) (Lepidoptera: Yponomeutidae) as a function of temperature: Effect on the number of generations in tropical and subtropical regions. NeotropEntomo., 41(5): 533-541.

Menéndez, R. 2007. How are insects responding to global warming? Tijdschriftvoor Entomologie., 150: 355-365.

Morimoto, N., Imura, O. and Kiura, T. 1998: Potential effects of global warming on the occurrence of Japanese pest insects – Applied Entomology and Zoology., 33(1): 147-155.

Parmesan, C., 2006. Ecological and evolutionary responses to recent climate change. Annu. Rev. Ecol. Evol. Syst. 37: 637-669.

Porter, J. H., Parry, M. L. and Carter, T.R. 1991. The potential effects of climatic change on agricultural insect pests. Agrl and Forest Meteorol, 221-240.

Scott, J. G., 1995. Effects of temperature on insecticide toxicity In: Roe, M. Kuhr, R.J. (Eds.), Reviews in Pesticide Toxicology, vol. 3. Toxicology Communications Inc., Raleigh, NC, pp. 111-135.

Sharma, H.C. 2010., Effect of climate change on IPM in grain legumes. In 5th International Food Legumes Research Conference (IFLRC V) and the 7th European Conference on Grain Legumes (AEP VII), 26–30th April 2010, Anatalaya, Turkey.

Sharma, H.C. and Ortiz, R., 2000. Transgenics, pest management, and the environment. Curr.Sci., 79: 421-437.

SibylleStoeckli, Martin Hirschi, ChristophSpirig, PierluigiCalance, Mathias W. Rotach and Jorg Samletz., 2012.Impact of Climate change on Voltinism and Prospective Diapause Induction of a Global Insect – *Cydia pomonella* (L.). PLoS one. 7(4): e35723.

Sridhar, V. and Reddy, P.V.R. 2013. Use of Degree day and plant phenology: A reliable tool for prediction of insect pest activity under climate change conditions In: Singh HP, Rao NKS and Shivashankar KS (Eds.) Climate Resilient Horticulture: Adaptation and Mitigation strategies (pp. 287-294). Springer Publishes, New Delhi.

Sridhar, V. and P. V. R. Reddy., 2012. Use of Degree day and plant phenology: A reliable tool for prediction of insect pest activity under climate change conditions pp 285-290. In: Adaptation and Mitigation Stratergies for Climate Resilient Horticulture (Eds.) K.S. Shivashankara, P. Patil, G. Selvakumar and V. Sridhar pp. 285-290, Springer publications.

Sridhar, V., Vinesh, L.S., Jayashankar, M. and Venugopalan, R. 2016. CLIMEX based spatio-temporal analysis for predicting the number of generations of *Spodoptera litura* (Fabricius) (Lepidoptera: Noctuidae) under climate change scenario, The Bioscan, 11 (2): 871-878.

Sridhar, V., Verghese A., Vinesh L. S., Jayashankar, M. and Kamala Jayanthi P. D., 2014. CLIMEX simulated predictions of Oriental fruit fly, *Bactrocera dorsalis* (Hendel) (Diptera: Tephritidae) geographical distribution under climate change situations in India. Curr. Sci., 106 (12): 1702-1710.

Thomas, C.D., Cameron, A., Green, R. E., Bakkenes, M., Beaumont, L. J., Collingham, Y. C., Erasmus, B. F. N., de Siqueira, M. F, Grainger, A., Hannah, L., Hughes, L., Huntley, B., van Jaarsveld, A. S., Midgley, G. F., Miles, L., Ortega-Huerta, M. A., Peterson, A. T., Phillips, O. L., and Williams, S. E. 2004. Extinction risk from climate change. Nature, 427: 145-148.

Thomson, L. J., Macfadyen, S., Hoffmann, A.A. 2010. Predicting the effects of climate change on natural enemies of agricultural pests. Biol. Control 52, 296-306.

Tobin, P. C., Nagarkatti, S., Loeab, G. and Saunders, M.C., 2008. Historical and projected interactions between climate change and insect voltinism in a multivoltine species. Global Change Biol14: 951-957.

Visser, M.E., Both, C and Lamberchts, M.M., 2004. Global climate change leads to mistimed avian reproduction. Adv. Ecol. Res., 35: 89-110.

Wang, X.Y., Yang, Z.Q., Wub, H. and Gould, J. R., 2007. Effects of host size on the sex ratio, clutch size, and size of adult Spathiusagrili, an ectoparasitoid of emerald ash borer. Biol. Control, 44: 7-12.

Ware, G.W., 2000. The Pesticide Book, Fifth ed. Thompson Publication, Fresno, CA.

White, T. C. R. 1984. The abundance of invertebrate herbivores in relation to the availability of nitrogen in stressed food plants. Oecologia, 63: 90-105.

Willis, C. G., B. Ruhfel, R. B. Primack, A. J. Miller-Rushing, and C. C. Davis. 2008. Phylogenetic patterns of species loss in Thoreau's woods are driven by climate change. Proc. Natl. Acad. Sci., USA. 105: 17029-17033.

Wilson, R. J., D. Gutierrez, J. Gutierrez, D. Martinez, R. Aguado, and Montserrat, V. J. 2005. Changes to the elevational limits and extent of species ranges associated with climate change. Ecol. Lett., 8: 1138-1146.

Young, A.M., 1982. Population Biology of Tropical Insects. Plenum Press, New York, USA, 511 pp.

Zalucki, M.P and Furlong, M.J., 2008. Predicting outbreaks of a migratory pest: an analysis of DBM distribution and abundance revisited. In: Srinivasan, R., Shelton, A.M., Collins, H.L. (Eds.), The Sixth International Workshop on Management of the Diamondback Moth and Other Crucifer Insect Pests NakhonPathom, Thailand, pp. 8-14.

Zavala, J.A., Cast, C.L., DeLucia, E.H. and Berenbaum, M. R., 2008. Anthropogenic increase in carbon dioxide compromises plant defense against invasive insects. Proc. Natl. Acad. Sci. U.S.A. 105: 5129–5133.

5

Effect of Climate Change on the Pollinators and Productive Insects

J. Ashoka and A.G. Sreenivas

Introduction

Most climate scientists agree the main cause of the current global warming trend is human expansion of the "greenhouse effect, warming that result when the atmosphere traps heat radiating from Earth toward space. On Earth, human activities are changing the natural greenhouse. Over the last century the burning of fossil fuels like coal and oil has increased the concentration of atmospheric carbon dioxide (CO_2). This happens because the coal or oil burning process combines carbon with oxygen in the air to make CO_2. To a lesser extent, the clearing of land for agriculture, industry, and other human activities has increased concentrations of greenhouse gases. The gases *viz.*, Water vapour, Carbon dioxide (CO_2), Methane, Nitrous oxide and Chlorofluorocarbons (CFCs) contribute to the greenhouse effect. Since 1750, the average amount of energy coming from the sun either remained constant or increased slightly. If the warming were caused by a more active sun, then scientists would expect to see warmer temperatures in all layers of the atmosphere. Instead, they have observed a cooling in the upper atmosphere, and a warming at the surface and in the lower parts of the atmosphere. That's because greenhouse gases are trapping heat in the lower atmosphere. When climate change occurs; temperatures can increase dramatically and when temperature rises, many different changes can occur on Earth. For example, it can result in more floods, droughts, or intense rain, as well as more frequent and severe heat waves. Oceans and glaciers have also experienced some changes: oceans are warming and becoming more acidic, glaciers are melting, and sea levels are rising. As these changes frequently occur in future decades, they will likely present challenges to our society and environment.

Further, some crops and other plants may respond favorably to increased atmospheric CO_2, growing more vigorously and using water more efficiently. At the same time, higher temperatures and shifting climate patterns may change the areas where crops grow best and affect the makeup of natural plant communities. It was expected that as the pace of climate change increased, there would be some degree of loss in biodiversity; the amount would be dependent upon the severity of temperature increase and precipitation changes in different parts of the world. Temperature increase was linked to the release of carbon dioxide and other gases that strengthened the greenhouse effect. All else being equal, rising carbon dioxide concentrations – the main driver of climate change – could increase production of some crops, such as rice, soybean and wheat. However, the changing climate would affect the length and quality of the growing season and farmers could experience increased damage to their crops, caused by a rising intensity of droughts, flooding or fires.

Prof. Samuel Myers, an environmental health expert at Harvard University, Boston, opined that "from a health viewpoint, iron and zinc are hugely important" and two billion people already suffer iron and zinc deficiencies around the world. This causes serious harm in particular to developing babies and pregnant women, and currently causes the loss of 63 m life years annually.

The observation made in March 2014 (by the IPCC) indicated that the parts of Tosa Bay (a body of water adjacent to the Japanese main island of Shikoku) that were once dominated by forests of kelp and sea grasses, along with animals associated with that ecosystem, were slowly changing into coral reef ecosystems with the help of tropical fish that were migrating northward.

Impact of Climate Change on Insects

Insects are good indicators of current human driven climate change and such changes in climate and weather could profoundly affect the population dynamics and status of insect pests of crops. They have responded to warming in all predicted ways from changes in phenology and distribution to undergoing evolutionary changes. Insects are among the groups of organism most likely to be affected by climate change because climate has a strong direct influence on their development, reproduction and survival. Moreover, insects have short generation times and high reproductive rates, so they can more like to respond quicker to climate change than long-lived organisms, such as plants and vertebrates (Menendez, 2007). Increasing climatic variability reduced the level of parasitism of caterpillars, which in turn may increase the frequency and intensity of herbivory outbreaks (Stireman, *et al.*, 2005). Further, the climate change affects biological processes such as metabolism, nutrient requirements and biochemical activities. The behavioural response like migration and geographic distribution are changed. Reproductive physiology and breeding behaviour of animals are directly affected by photoperiods, temperature and precipitation. Insects relying on long-range signalling will be among the most impacted. The quantitative composition of complex pheromone blends will be impacted. Understanding how climate change will affect species interactions is a challenge for all branches of ecology. We have only limited understanding of how increasing temperature and atmospheric CO_2 and ozone levels will affect pheromone-mediated communication among insects. The abiotic parameters are known to have direct impact on insect population dynamics through modulation of developmental rates, survival, fecundity, voltinism and dispersal. Among the climatic factors, temperature is an important factor and the alteration in the voltinism also could be the results of warming. It is more profit to multivoltine species and voltinism could be reflected in changes in the geographical distribution.

Bumblebees are important pollinators of wild plants as well as of food crops. Other research has shown that the common Eastern bumblebee, for example, provides pollination services worth an average of $390 per acre of crops in areas where it is found. As a result, bumblebee decline could have serious implications for both ecosystem function and food production. Bumblebees already have to contend with lethal infections, deadly pesticides and habitat loss due to agriculture and urban development. A new study shows they have another cause for worry: climate change; rising temperatures due to climate change are latest threat to bumblebees. If humans continue to warm the climate, "it doesn't look very good for bumblebees." Cool-adapted bumblebees are failing to shift their ranges in response to climate warming. The study also reported that across both Europe and North America, the southern ranges for bumblebees moved northward by an average of 185 miles between 1901 and 2010. Bumblebees are especially vulnerable to extreme heat events, which have become more frequent as a result of climate change and the bees did not adapt by moving farther north, as butterflies and some other species have done, according to Jeremy Kerr.

Figure 1: Rising Temperatures due to Climate Change are Latest Threat to Bumblebees.

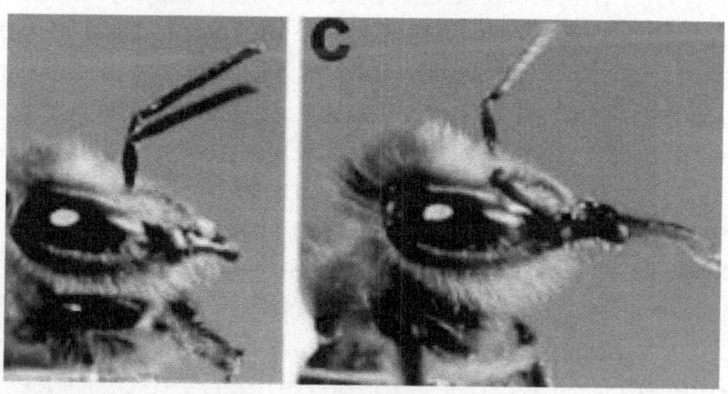

Figure 2: Honeybees can't Smell Flowers well Amid Pollution

Jeremy Kerr analyzed data on 67 bumblebee species in Europe and North America, examining changes in their northern and southern range limits. They found that bumblebees on both continents were experiencing dramatic losses in the southern parts of their ranges. In other words, climate change has been driving the southern edge of the bumblebees' range northward.

Pollination activities, out of total over 80 per cent is performed by insects (Robinson and Morse 1989), is negatively impacted by climate change which result on lower production of crop may lead to serious global food security problem. Many experiments proved that increased temperature has effect on plants, pollinators and their interactions. Changing climates may cause changes in the time of growth, flowering and maturation of crops, with consequent impacts on crop-associated biodiversity, particularly on pollinators. Key biological events such as insect emergence, their foraging behaviour and date of onset of flowering need to occur in synchrony for successful pollination interactions. On a larger scale, changes in temperature, disturbances on rainfall pattern and other many environmental changes over the entire season may alter the abundance, diversity and foraging behaviour of pollinators. Hence climate change causes very serious impact on insect pollinators and flowering plants.

Significant variation exists in the physiological, biochemical and molecular responses to elevated CO_2 atmosphere, among terrestrial plant species including those with C_3, C_4 and crassulacean acid metabolic (CAM) pathways. The growth parameters of mulberry *viz.*, plant height, and the number of leaves, the leaf yield and total plant biomass were higher in the elevated climate change treatments as compared to ambient treatments. Growth indices, especially leaf area index (LAI), nitrogen based index (NBI) values were highest in eCO_2 alone and in combination with temperature. The direct effects of eCO_2 on individual plant species have been well documented (Curtis and Wang, 1998) wherein, eCO_2 generally resulted in increased leaf area index and increased growth rate. The plant physiological and biochemical constituents *viz.*, chlorophyll, tannin, phenol, carbohydrates, carbon, total sugars, crude fibre showed significant increase while, flavonoids, nitrogen and proteins exhibited decreased values in eCO_2 conditions.

Further, biological studies on silkworm, *Bombyx mori* L. on mulberry (grown under different climate change treatments) was conducted during 2015-16 and the results indicated that there was decreased larval weight, increased larval duration, decreased effective rearing rate in eCO_2 and temperature conditions. The change in phytochemistry of mulberry plant due to eCO_2 and temperature has influenced the changes in silkworm growth and development. This was further evidenced by the morphometric measurements of larva; the average length, breadth, head capsule breadth of larvae varied in different treatments. The silkworm larvae fed with mulberry which was grown under ambient atmosphere (reference plot) registered significantly higher length, breadth and head capsule measurements than elevated treatments. It is evident from the current studies that the size of the larvae was smaller in eCO_2 treatments than in a CO_2 treatments. The temperature and CO_2 prevailing in reference plot supported the mulberry to produce more and acquiring more nutritional quality parameters and that in turn nurtured the silkworm larva for better survival and acquiring body mass to yield more in terms of cocoon, pupal, shell and filament weights and filament length when compared to other elevated CO_2 treatments. The higher values obtained for the above traits may be attributed to the higher nutrition obtained by the silkworms from the mulberry leaves grown in a CO_2 atmosphere conditions. The higher pupal weight, higher moth emergence, higher fecundity and shorter pupal duration was observed for *B. mori* in reference (ambient conditions) treatment. The study revealed that the silkworm larvae fed with the leaves of reference plot recorded higher fecundity compared to elevated treatments. Earlier studies by Milam *et al.* (2011) reported that the plants grown under elevated conditions when fed to larvae resulted in lower relative growth rates for the larvae as the needed N concentration and water content were lower in elevated CO_2 treatments. Lower growth rates resulted in lower body mass in elevated CO_2 treatments and as a result fewer number of eggs per adult moth and vice-versa. Further, when Japanese beetle (*Popillia japonica*) fed foliage developed under elevated CO_2, the lifespan of beetles was prolonged by eight days with reduced fecundity as reported by Bridget *et al.* (2008) and this analogy holds good for silkworm larval durations, time to moth emergence and fecundity.

Differences between cultivars of a single crop suggest that breeding for decreased sensitivity to atmospheric CO_2 concentration could partly address these new challenges to global health. Efforts are needed to address climate change reversal and take steps to slow down climate change to prolong the comfort of earths' atmosphere for all living forms. From the point of productive insects, the enhancement of quality of foliage on which the productive insect depend is important and such studies needs to be initiated as a part of mitigation strategies.

References

Bridget, F. O., Arthur, R. Z., Evan, H. D. and May, R. B., 2008, Longevity and fecundity of Japanese beetle (*Popillia japonica*) on foliage grown under elevated Carbon dioxide. *Environ. Entomol.*, 37(2): 601-607.

Curtis, P. S. and Wang, X., 1998, A meta-analysis of elevated CO_2 effects on woody plant mass, form and physiology, Oecol., 113: 299-313.

Jeremy T. Kerr, Alana Pindar, Paul Galpern, Laurence Packer, Simon G. Potts, Stuart M. Roberts, Pierre Rasmont, Oliver Schweiger, Sheila R. Colla, Leif L. Richardson, David L. Wagner, Lawrence F. Gall, Derek S. Sikes, Alberto Pantoja (2015), Climate change impacts on bumblebees converge across continents. Science: 177-180.

Menendez, R. 2007. How are insects responding to global warming. *Tijdschrift voor Entomologie* 150: 355-365.

Milam, E. S., Micheal, A. D., Seth, G. P., Brett, R. G., Stephen, A. P., Hank, E. S., Hugo, H. and Roland, R. D., 2011, Influence of elevated CO_2, nitrogen and Pinuselliottii genotypes on performance of the redheaded pine sawfly, Neodiprion lecontei. Canadian., J. Forest. Res., 34(5): 1007-1017.

Robinson W E, Morse R A., (1989), The value of honeybees as pollinators of US crops. American Bee Journal, 129 (1): 477-487.

Samuel S. Myers, Antonella Zanobetti, Itai Kloog, Peter Huybers, Andrew D. B.Leakey, Arnold J. Bloom, Eli Carlisle, Lee H. Dietterich, Glenn Fitzgerald, Toshihiro Hasegawa, N. Michele Holbrook, Randall L. Nelson, Michael J. Ottman, Victor Raboy, Hidemitsu Sakai, Karla A. Sartor, Joel Schwartz, Saman Seneweera, Michael Tausz and Yasuhiro Usui, (2014), Impacts of elevated atmospheric CO_2 on nutrient content of important food crops. Nature, 510, 139–142.

Stireman, J. O.; Dyer, L. A.; Janzen, D. H. and Singer, M. S. (2005). Climatic unpredictability and parasitism of caterpillars: implications of global warming. *Proceedings of the National Academy of Science*. 102: 17384-17387.

6

Effect of Climate Change on Natural Enemies Complex of Crop Pests

Arun Kumar Hosamani, Mallikarjun Warad, G. Akshatha, Vishal Kumar and Ambuja

Introduction

Global climate change is an issue of global concern. It is evident from increase in global average temperature, changes in the rainfall pattern and extreme climatic events. The global CO_2 concentration, has increased over approximately 25- 30 % during the last 100-120 years. Similarly major greenhouse gases (CH_4, NO_2 and carbohalogenides) concentration is also increasing in the environment with various natural and anthropogenic activities as well (Vijaya Venkata Raman *et al.*, 2012). This will lead to a rise in global temperature. The global temperature is rising at a rate of 0.1-0.8°C per decade. Current estimates of changes in climate may leads to an increase in global mean annual temperatures of 1°C by 2025 and 3°C by the end of the next century (IPCC, 1990). In Agriculture, biotic and abiotic stresses influence crop growth and productivity to the extent of 80%. These yield losses are likely to be aggravated with to climate change. Among biotic stresses, insects are major organisms. Insects are poikilothermic (cold blooded animals), they change their body temperature according to its surrounding temperature. Hence, global warming can majorly influence the population dynamics of insect pests and also the activity and abundance of natural enemies.

Climate change will have indirect influence on natural enemies via insect pests and host plants. As CO_2 level gets elevated plants respond to it with higher biomass. As a result there is decrease in nitrogen level in plants. Such plants become nutritionally poor to herbivores and also to their natural enemies. Increased temperatures leads to asynchronization between the natural enemies and their hosts thus affect the effectiveness of their natural control. Host-specific Hymenopteran parasitoids are greatly affected than the general feeders.

The ability of natural enemies to attack their hosts will depend on their tolerance to environmental extremes. As climate becomes more variable, the pest outbreaks will increase and disrupt the ability of existing natural enemies. Under the influence of climate change, the effectiveness of some biological control agents may change. These changes will have major impact on biological control, particularly in the developing countries. Keeping these facts in view the topic, the potential impacts of climate change on bio-control agents is discussed.

Effect of Climate Change on Insect Pests

The climatic parameters will have direct impact on insect pests. Change in climate leads to increase in the rate of development, feeding, survival, fecundity and dispersal abilities of the insect pests. Among the climatic factors, temperature is an important. Warmer conditions are likely to increase the outbreak of pests (Cammell and Knight, 1992).

Temperature will have a profound effect on the relationship of pests and predators. Natural enemy and host insect populations will respond differently to changes in temperatures. There will be lack of synchrony between the vulnerable stage of the pest and its natural enemy, particularly with parasitoids. Hosts may pass though vulnerable life stages more quickly at higher temperature, reducing the window of opportunity for parasitism. Rising temperatures might have a negative effect on delicate natural enemies such as Hymenopteran parasitoids and small predators. Abdul (2017) found that elevated CO_2 has increased the reproductive parameters of larval parasitoid *Campoletis chloridae* of *Helicoverpa armigera* in chickpea.

Effects of Climate Change on Natural Enemies

Survival and Developmental rate: Natural enemies will have certain optimum temperature preference below and above which their survival and developmental rate is affected. Larval endoparasitoid, Campoletis chlorideae of Helicoverpa armigera did not survived when larvae reared at <12 and >35°C after parasitisation. It suggests that temperatures above 35°C as a result of global warming will be lethal for development and survival of immature stages of °C chlorideae. Developmental rate is negatively correlated with the temperature. The parasitoid adults stored at 12°C survived for longer period of time. The efficiency of parasitism and adult emergence were quite low (Dhillon and Sharma, 2009).

Feeding potential: The fitness of predators and parasitoids increases with their herbivore hosts quality. The coccinellid predator, *Coccinella sexmaculata* of *Bt* cotton aphid consumed significantly more prey under conditions of elevated CO_2 and temperatures as compared to ambient conditions (Shreevani, 2015). Therefore bio-control efficiency of the coccinellids against aphid hosts may be enhanced in elevated CO_2 and temperatures. These results suggest that higher CO_2 could make generalist predators more effective in controlling pests.

Prey searching ability: The searching time needed by natural enemies to locate their prey could be increased as a consequence of extra plant foliage due to increased plant growth. Plants growing under increased CO_2, temperature extremes and decreased precipitation will provide different nutritional resources for insect pests, indirectly influencing the fitness of predators and parasitoids feeding on these hosts.

Physiology of natural enemies: Climate change will have direct influence on the physiology of insect natural enemies. The increased temperature affects the immune response of insect pests to parasitoid attack, thus it have negative impact on parasitoid efficacy. Thereby affects their ability to control insect pests (Stacey and Fellowes, 2002). Natural enemies also influenced by warmer conditions in similar ways to that of insect pests, with faster growth rates, smaller body sizes and reduced longevity (Saona and Miller, 1999).

Shift in geographical area: Modifications in the geographical distribution of plants, hosts and parasitoids may result from a number of processes. With increase of 2 and 3°C in average daily temperatures, the geographical area favourable for mealy bug development increases. It is due to increasing failure of biological control. Cold-limited species may be able to increase their geographic range in the future. For example, global warming will allow the cold intolerant pink bollworm, Pectinophora gossypiella (Saunders) (Lepidoptera: Gelechiidae), to expand its range on Bt cotton (Gutierrez *et al.*, 2008).

Synchronization of host and natural enemy: Climate change may cause temporal asynchrony between pest and its natural enemy, thus reduce parasitism (Karuppaiah and Sujayanad, 2012). The parasitoid *Campoletis chlorideae* developed successfully over the temperature range of 12-37 °C. However, the developmental period was found to be inversely correlated with temperature in the range of 12-37 °C. As temperature increased, longevity decreased accordingly (Pandey and Tripathi, 2008).

Diapause: Changes in day length and temperature as a result of climate change prevents the diapause induction in natural enemies.

Change in climate can differentially affect the developmental rates of pest and natural enemy population. Hence, in biological control system of insect pests utilising parasitoids, the temperature response of the parasitoids determines their success in controlling the pest population. Parasitoids often have a lower temperature tolerance than their hosts, they must be able to overwinter to survive. The egg predator *Cyrtorhinus lividipennis* of brown plant hopper (BPH) had increased instantaneous attack rates and decreased handling times with increasing temperatures until 32°C. At 35°C, the attack rate and handling time decreased drastically. This implies that predator activity is likely to increase with increasing temperatures up to a critical temperature of about 35°C (Karuppaiah and Sujayanad, 2012).

Furthermore, the natural enemies of insect pests interact with plants, even if they do not rely on them directly for their survival. Many natural enemies use plant chemical signals to locate their prey. Higher temperatures and elevated levels of CO_2 could interfere with host location finding (Stiling *et al.*, 2002). For example, *Diaeretiella rapae* parasitoids were shown to be attracted to mustard oils from the host plant. Their prey aphid *Myzus persicae* have mustard oil content in their body. Warmer temperature increases the volatility of these chemicals and should aid the parasitoids in finding their hosts. Therefore changes in plant biology will act on the natural enemies as well as the herbivores. Increase in the atmospheric CO_2, changes in precipitation and increased temperature will alter plant phenology, influencing herbivore growth and abundance, and thus indirectly affecting the natural enemies.

References

Abdul Biradar, 2017, Ph. D Thesis, Submited to University of Agricultural Sciences, Raichur.

Cammell, M. and Knight, J., 1992, Effects of climate change on the population dynamics of crop pests. *Adv. Ecol. Res.*, 22: 117-163.

Dhillon, M. K. and Sharma, H. C., 2009, Temperature influences the performance and effectiveness of field and laboratory strains of the ichneumonid parasitoid, *Campoletis chlorideae. Bio Control.*, 54: 743-750.

Gutierrez, A. P., Daane, K. M., Ponti, L., Walton, V. M. and Ellis, C. K., 2008, Prospective evaluation of the biological control of vine mealy bug: refuge effects and climate. *J. Appl. Ecol.*, 45(2): 524-536.

IPCC, 1990, The potential impacts of climate change on agriculture and forestry. Inter governmental panel on climate change. Geneva and Nairobi, Kenya: World Meteorological Organization and UN Environment Program. p. 55.

Karuppaiah, V. and Sujayanad, G. K., 2012, Impact of climate change on population dynamics of insect pests. *World J. Agrl. Sci.*, 8(3): 240-246.

Pandey AK, Tripathi CPM (2008). Effect of temperature on the development, fecundity, progeny sex ratio and life-table of Campoletis chlorideae, an endo-larval parasitoid of the pod borer, Helicoverpa armigera. Bio Control. 53: 461-471.

Saona RC, Miller JC (1999). Temperature-dependent effects on development, mortality, and growth of *Hippodamia convergens* (Coleoptera: Coccinellidae). Environ. Entomol. 28: 518.

Shreevani, G. N., 2015, *Ph. D Thesis*, Submited to University of Agricultural Sciences, Raichur.

Stacey DA, Fellowes MDE (2002). Elevated CO_2 effects on three trophic levels: how plants, aphids and their natural enemies respond. Global Change Biol. 8: 668- 678.

Stiling P, Cattell M, Moon DC, Rossi A, Hungate BA, Hymus G, Drake B (2002). Elevated atmospheric CO_2 lowers herbivore abundance, but increases leaf abscission rates. Global Change Biol. 8: 658-667.

Vijaya Venkata Raman S, Iniyanb S, Goicc R (2012). A review of climate change, mitigation and adaptation. *Renewable and Sustain. Energy Rev.* 16: 878- 897.

7

Emergence of Plant Diseases Due to Climate Change and their Management

Gururaj Sunkad, Y.S. Amaresh, D.S. Aswathanarayana, A.S. Savitha and T.H. Shruthi

Introduction

Increasing world food production by more than 50 % is the projected demand of the exponential growing of world's population by 2050. Achievement of this target is harder due to drastic climatic changes which affect the world food production. In the last four decades, management of pest and diseases have played its role in doubling food production, but fungal, bacterial and viral pathogens still claim 10 % - 16 % of the global agricultural products harvest. The impact of climate change on the complex biological interactions very much affects pests and pathogens which eventually results in affecting the grain production. Currently, manipulations of this mitigate climatic change impact is the big challenge being faced by the agricultural scientists.

Human activities are altering greenhouse gas concentrations in the atmosphere and causing global climate change. The impacts of climate change can be positive, negative or neutral, since these changes can decrease, increase or have no impact on diseases, depending on each region or period. These impacts will also be observed on plants and other organisms as well as on other agroecosystem components. From disease management or protection point of view we have to prepare ourselves and have to take diverse of measures in collaborations with epidemiologist, climatologist, sociologist, economist, Geological Information system (GIS) specialist and management practitioners to have a better understanding of plant pathogen interactions under climate change conditions.

Greenhouse gas concentrations in the atmosphere are being altered by human activities, thus causing global climate change. These activities, intensified after the industrial revolution at the end of the eighteenth century result from the use of natural resources such as fossil fuel burning, deforestation and other land use changes. The atmospheric concentration of carbon dioxide (CO_2) has reached levels significantly higher than in the last 650 thousand years (Siegenthaler *et al.*, 2005). Since 2000, the growth rate in CO_2 concentration is increasing more rapidly than in the previous decades (Canadell *et al.*, 2007). Similar trends have been observed for methane (CH_4), nitrous oxide (N_2O), and other greenhouse gases (Spahni *et al.*, 2005; IPCC, 2007). Parameters like elevated temperatures, carbon dioxide concentration, and rainfall pattern can influence the development of both hosts, pathogen which ultimately, impact disease development.

Due to different diseases, insects and weeds a total of 36.50 % loss was estimated and 14.10 % of crop losses are attributed due to the occurrence of diseases, which represents an annual loss of $220 billion dollars (Agrios, 2005). Diseases are responsible for losses of atleast 10 per cent of global food production, representing a threat to food security (Strange and Scott, 2005). Besides direct losses, the methods for disease control especially the chemical methods can result in environmental contamination and in residual chemicals in food, in addition to social and economic problems. The close relationship between the environment and diseases suggests that climate change will cause modifications in the current phytosanitary scenario. The impacts can be positive, negative or neutral, since there can be a decrease, an increase or no effect on the different pathosystems, in each region. The analysis of the potential impacts of climate change on plant diseases is essential for the adoption of adaptation measures, as well as for the development of resistant cultivars, new control methods or adapted techniques, in order to avoid more serious losses (Chakraborty and Pangga, 2004; Ghini, 2005).

Up to the end of last century, few data report the impact of climate change on plant diseases and in most cases refer only to the increase of some pollutants (mainly ozone, SO_2 and CO_2) in the atmosphere and their direct effects on plants. Ozone and CO_2 concentrations have increased since the industrial revolution; however, they can alter plant physiology in opposite ways. Plants are subjected to acute and chronic exposure of ground-level ozone resulting in symptoms of foliar injury on sensitive plants. In addition, many physiological functions necessary for growth and reproduction are impaired by O_3. In addition, photosynthesis, leaf area, plant height and crop yield *etc.* are increased in the presence of higher doses of CO_2 (Eastburn *et al.*, 2011). Lastly, elevated CO_2 levels tend to reduce damage caused by high ozone concentrations (Tiedemann and Firsching, 2000). Global climate change, particularly increased in temperature and CO_2 levels (IPCC, 2007) are the consideration to manipulate all the disease triangle elements *i.e.*, host, pathogen and climate factors and their interactions (Legreve and Duveiller, 2010). Thus, climate change may influence disease incidence and severity and also influence further coevolution of plants and their pathogens (Chakraborty, 2005; Burdon *et al.*, 2006; Eastburn *et al.*, 2011).

Once environment and diseases are closely related, climate change will alter the geographical and temporal distribution of phytosanitary problems. Pathogens and other microorganisms related to the disease process will be affected. Therefore, new diseases may arise in certain regions, and other diseases may cease to be economically important, especially if the host plant migrates into new areas (Coakley *et al.*, 1999). Chakraborty *et al.* (2000) reported more aggressive strains of pathogen with broad host range, such as *Rhizoctonia, Sclerotium* and other necrotrophic pathogens can migrate from agroecosystems to natural vegetation and less aggressive pathogens from natural plant communities can start causing damage in monocultures of nearby regions. Overall, climate change is predicted to have a progressively negative effect on the yield of food crops, particularly in the absence of efforts to mitigate global CO_2 emissions. As with all species, plant pathogens will have varying responses to climate change. Whilst the life cycle of some pathogens will be limited by increasing temperatures, *e.g. Puccinia striiformis* f. sp. *tritici*, other climatic factors such as increasing atmospheric CO_2, may provide more favourable conditions for pathogens such as *Fusarium pseudograminearum*.

Climate change directly impacts crops, as well as their interactions with microbial pests (Rosenzweig *et al.*, 2000). However, little work has been done to model the effects of climate change on plant disease epidemics (Garrett *et al.*, 2006). Changing weather can induce severe plant disease epidemics (Chakraborty, 2005; Bosch *et al.*, 2007), which threaten food security if they affect staple crops (Anderson *et al.*, 2004) and can damage landscapes if they affect amenity species (Bergot *et al.*, 2004). There is a paradigm shift in nature, time and type of occurrence of viral and other diseases of various horticultural crops due to climate change.

Commonly, necrotrophic and biotrophic pathogens show a different behavior as regards their nutrition. The necrotrophic obtain nutrients from dead tissues, while the biotrophic derive nutrients from living cells and maintain a prolonged and deep interaction with their host. Therefore, all climatic factors (high temperatures or O_3 levels) that cause or accelerate tissue death could favour necrotrophic pathogens. In addition, climatic factors that stimulate plant growth such as elevated levels of CO_2 or increased temperature will change the physiology of the plant, altering host colonization by biotrophic pathogens (Elad and Pertot, 2014). In particular, the decomposition of high CO_2 litter occurs at a slow rate, increasing the biomass availability which combined with higher winter temperatures, might increase pathogen survival on crop residues accumulating the amount of initial inoculum to infect subsequent crops (Coakley *et al.*, 1999).

One of the first studies about climate change on plant diseases was conducted by Luo *et al.* (1995). A risk analysis of rice blast epidemics and plant growth associated with climate change in several Asian countries due to the importance of this crop and to the losses related to this disease, caused by *Magnaporthe grisea* were reported. Simulations risk of blast epidemics under the effects of temperature change and enhanced UV-B radiation were studied. Changes in the amount of rainfall, do not affect the occurrence of the epidemics since they have little effect on the leaf wetting period. In cool subtropical zones, higher temperatures caused increases in disease severity and in the area below the disease progress curve, because higher risk of epidemics occurs under higher temperatures. In humid tropical and humid warm subtropical zones, such as Southern China, Philippines and Thailand, the opposite effect was observed. Lower temperatures increased the risk of rice blast epidemics since the current temperatures in these regions are above favorable values for the occurrence of this disease. However, a larger area below the disease progress curve does not always result in lower rice yield, since the effect in plant growth also takes place. The effects of the increase in UV-B radiation were highly significant for the occurrence of epidemics.

Increased CO_2 levels can impact both the host and the pathogen in multiple ways. New races may evolve rapidly under elevated temperature and CO_2, as evolutionary forces act on massive pathogen populations boosted by a combination of increased fecundity and infection cycles under favourable microclimate within enlarged canopy (Chakraborty, 2013). Alternatively, a high concentration of carbohydrates in the host tissue promotes the development of biotrophic fungi such as rust (Chakraborty *et al.*, 2003). Thus, an increase in biomass can modify the microclimate and affect the risk of infection (Lambers *et al.*, 2008). In general, increased plant density will tend to increase leaf surface wetness duration and regulate temperature, and so make infection by foliar pathogens more likely (Dalla Pria *et al.*, 2006).

Elevated levels of CO_2 can directly affect the growth of pathogens, e. g., according to Chakraborty *et al.* (2003), the growth of the germ tube, appressorium and conidium of *Colletotrichum gloeosporioides* fungi is slower at high concentrations of CO_2 (700 ppm). Germination rates of conidia on leaves were lower at CO_2 concentrations of 700 ppm than those observed at 350 ppm. However, once the pathogen infects the plant, the fungus quickly develops and achieves sporulation. In contrast, the rate of germination and sporulation was greater at high concentrations of CO_2 (700 ppm). Elevated ozone can have a similar effects such as a 3 to 5 fold increase in rust infection on poplar, however this response is reduced by elevated CO_2 (Karnosky *et al.*, 2002).

A change in temperature may favor the development of different inactive pathogens, which could induce an epidemic. Due to changes in temperature and precipitation regimes, climate change may alter the growth stage, development rate, pathogenicity of infectious agents, and the physiology and resistance of the host plant (Charkraborty and Datta, 2003). Evidence suggests that sunlight affects pathogens due to the accumulation of phytoalexins or protective pigments in host tissue. Temperature is one of the most important factors affecting the occurrence of bacterial diseases such as *Ralstonia solanacearum*, *Acidovorax avenae* and *Burkholderia glumea*. Thus, bacteria could proliferate in areas where temperature dependent diseases have not been previously observed (Kudela, 2009). As the temperature increases, the duration of winter and the rate of growth and reproduction of pathogens may be modified (Ladanyi and Horvath, 2010).

Among the factors responsible for the increased importance of disease, the possibility of its dissemination to new areas, a longer favorable period for inoculum production and infection, higher dissemination and survival rates in roots and the soil, besides an increased plant predisposition. A similar study was conducted by

Booth *et al.* (2000) who performed a risk analysis for the occurrence of *Cylindrocladium quinqueseptatum* in eucalyptus in the various production regions in the world.

All of these pioneering studies assumed constant increments in temperature, precipitation, or other climate variables. However, the development and availability of global climate models (General Circulation Models, GCM) enabled the application of these spatial increments in the risk evaluation of disease occurrence. One of the most important accomplishments of the Intergovernmental Panel on Climate Change (IPCC) was the presentation of a group of greenhouse gas emission scenarios known as SRES (Special Report on Emissions Scenarios), which take into account driving forces such as population, economy, technology, energy, agriculture (land-use), as well as their interactions. Thus, four scenario families were defined: A_1, A_2, B_1 and B_2, which constitute references related to future greenhouse gas emissions (IPCC, 2001).The IPCC Data Distribution Centre (DDC) was established to facilitate the dissemination of these scenarios, as well as of related environmental and socio-economic factors, for climate impact evaluations.

Models can simulate global climate change scenarios and different levels of disease severity, with the aim to estimate yields, establish control tactics and strategies, aiding the decision making process. More efforts must be dedicated to this type of study, which can result in significant time and resource savings. Lopes *et al.* (2008) studied, if the predictions about a rise in global temperature actually take place, rainy season potato crops will be more stricken with disease than winter crops, and cropping will not be feasible in low altitude regions due to the increase of biotic and abiotic diseases.

The influence of climate change on the biology of the host is often underestimated: phenology, sugar, acid, nitrogen and phenolic contents but also the number and size of leaves, stomatal density amount and composition of wax on leaves can be considerably modified by climatic conditions. Moreover, two main factors influencing a correct development of crop plants, such as prolonged dormant period during winter (chilling requirement) and heat unit accumulation, could be altered by higher temperatures in the winter resulting in an insufficient chilling period with consequent decreasing of bud breaks and prolonged flowering period. Otherwise, the increased temperatures in winter and spring can anticipate the maturation of ascospores and their release, forcing early starting of disease management. The general increase in temperature also produces an extension of the vegetative season, exposing crops to a higher number of secondary infections.

Gour *et al.* (2000) observed the luxuriant growth and toxin production by *Xanthomonas axonopodis* pv. *vignicola*, causal organism of leaf blight of cowpea was recorded at a temperature of 30°C and at pH of 7.0. Growth declined considerably at pH values higher and lower than 7.0, being minimum at pH 5.0 and 9.0. *Xanthomonas axonopodis* pv. *punicae*, developed at wide temperature range of 20 to 40°C and documented maximum number of colonies, when inoculated plates were incubated at 27°C. Similarly, pH of 7.2 was found optimum with a range between 5.5 - 8.0.

Pangga *et al.* (2004) stressed the importance of induced resistance studies, besides canopy size, in the evaluation of CO_2 effects. The reduction in incidence and severity of the red maple fungal pathogen (*Phyllosticta minima*) in *Acer rubum* in a CO_2 enriched atmosphere was a consequence of stomatal conductance reduction providing smaller openings for infecting germ tubes (Mcelrone *et al.*, 2010). The reduction in silicon content in rice leaves due to a reduction in respiration rates was the explanation for an increase in rice blast severity in a FACE (Free Air Carbon Dioxide Enrichment) experiment (Kobayashi *et al.*, 2006). According to Braga *et al.* (2006), the exposure to CO_2 enriched atmospheres can change inducible defensive responses in plants against pathogens. These changes occurred in individual metabolites and were dependent on cultivar resistance patterns.

Gioria *et al.* (2008) showed a prediction on main tomato diseases and argued that climate change will not be favorable for the occurrence of late blight (*Phytophthora infestans*), verticillium wilt (*Verticillium albo-atrum*), and white mold (*Sclerotinia sclerotiorum*); and will not alter the importance of the tomato mosaic caused by the tomato mosaic virus (ToMV) and the *septoria* leaf spot (*Septoria lycopersici*). In contrast to those diseases, the importance of powdery mildew (*Leveilula taurica*) will increase in all tomato production regions across the country, just as the importance of early blight (*Alternaria solani*), *fusarium* wilt (*Fusarium oxysporum* f. sp. *lycopersici*), bacterial wilt (*Ralstonia solanacearum*), tomato spotted wilt virus (TSWV), tomato chlorotic spot virus - TCSV, groundnut ring spot virus-(GRSV), Chrysanthemum stem necrosis virus - CSNV and yellow leaf curl virus (Geminivirus).

During 2006, higher incidence of sunflower powdery mildew ranging from 70 to 100 % observed in different parts of Karnataka. From then the disease is regularly seen in different parts of Karnataka in a moderate to severe form. Sudden outbreak of this disease is one among the varied reasons for significant reduction in area under sunflower cultivation.

Effect of elevated CO_2 and inter annual climatic variability affect *Cercospora* leaf spot diseases of two deciduous trees. When significant changes occur, disease incidence and severity always increased under elevated CO_2 (Mcelrone *et al.*, 2010). *Phytophthora* sp. which increased the incidence of root rot in central Europe forest trees as a result of the rise in mean winter temperatures, the shift in precipitation from summer to winter and the tendency toward heavier rains (Elad and Pertot, 2014).

In addition, strong effects of temperature on pathogen fitness traits as well as genotype x temperature interactions were found for *Puccinia striiformis* f. sp. *tritici*, the agent of wheat yellow/stripe rust. There are indications of increased aggressiveness of stripe rust isolates adapted to higher temperatures, suggesting that rust fungi can adapt and benefit from warmer climate conditions (Mboup *et al.*, 2012).

A critical analysis of the weather data (2000-2010) of the major chickpea growing areas in India indicated higher incidence of dry root rot disease (DRR) caused by *Rhizoctonia bataticola* (*Macrophomina phaseolina*) of chickpea in years when temperature exceeds 33°C. Increasing incidence of DRR at various locations over years suggest a strong influence of climate change on disease infection. Savary *et al.* (2011) categorize DRR as an acute-emerging disease that occurs irregularly, both temporally and spatially, may cause massive disruptions in system performances and whose range is expanding to new areas. Mamta Sharma and Suresh Pande

(2013) conducted studies to quantify the role of temperature and soil moisture associated with infection, colonization and development of DRR under controlled environment. The DRR incidence was significantly affected by high temperature and soil moisture deficit. Out of five temperature regimes (15°C, 20°C, 25°C, 30°C and 35°C) and four moisture levels (40 %, 60 %, 80 % and 100 %), a combination of high temperature (35°C) and soil moisture content (60 %) predisposes chickpea to DRR. The study clearly demonstrates that high temperature coupled with soil moisture deficit is the climate change variables predisposing chickpea to *R. bataticola* infection, colonization and development.

Leptosphaeria maculans and *L. biglobosa* are closely related sibling fungal pathogens that cause phoma leaf spotting, stem canker (blackleg) and stem necrosis of oilseed rape (*Brassica napus*). Information about the time of ascospore release at a particular location provides important data for decision making in plant protection, thereby enabling fungicides to be used only when necessary and at optimal times and doses. Although the pathogens have been studied very extensively, the effect of climate change on the frequencies and distributions of their aerially dispersed primary inoculum has not been reported to date. The average air temperature and precipitation for the time of development of pseudothecia and ascospore release (July-November), increased during the years under study at the rates of 0.1 °C and 6.3 mm per year. The day of the year (DOY) for the first detection of spores, as well as the date with maximum of spores, shifted from 270 to 248 DOY, and from 315 to 265 DOY, respectively. The acceleration of the former parameter by 22 days and the latter by 50 days has great influence on the severity of stem canker of oilseed rape (Joanna *et al.*, 2016).

Rice blast caused by *Magnaporthe oryzae* (Herb.) is very serious and recurrent problem in all rice-growing regions of the world. In India, rice blast is a serious concern due to favorable climatic condition during the crop season. Temperature rise due to climate change is expected to change pathogenicity of the pathogen. Temperature has a significant influence on growth and sporulation of rice leaf blast pathogen. Both growth and sporulation were increased up to a temperature (27°C) and declined further in response to increased or decreased with temperature (32°C and 22°C). Rise in temperature leads to increase in growth and sporulation of *M. oryzae* in tropical and subtropical regions of the world, it's to be expected to enhance in aggressiveness of *M. oryzae* that results rice blast epidemic in tropical and subtropical regions of the world. Therefore, the impact of temperature on infection components may use for development of crop loss assessment models, rice blast prediction models, evaluation of genotypes for the development of future plant diseases managements strategies (Laxman *et al.*, 2017).

Angelotti *et al.* (2017) studied potential impact of climate change on the occurrence of grapevine downy mildew in Brazil. Seedlings containing four to six leaves were sprayed with a sporangia suspension containing 10^5 sporangia per milliliter. After spraying, the seedlings were subjected to temperatures of 26, 28, 29.1, 30.4, and 31.8°C for 24 hours. The per cent of diseased leaf area and the latent period were evaluated. Maps of the geographic and temporal distribution of the disease were made considering the monthly average of the mean air temperature and leaf wetness duration for the reference climate or climate normal (1961-1990) and the future climates (2011-2040, 2041-2070, and 2071-2100), considering the A_2 and B_1 gas emission scenarios, designed by the Inter-governmental Panel on Climate Change (IPCC). Favorability ranges were set and used in logic functions of the geographical information system (GIS) to generate monthly maps for grapevine downy mildew. Rising temperatures interfered with the grapevine downy mildew infections, reduced the disease severity, and increased the latent period. Future climate scenarios indicate a reduction of favorability of downy mildew in Brazil, with variability in the different grape producing regions.

Beneficial fungi that infect plant roots and assist in absorption of nutrients, known as mycorrhizae, although not exposed to solar radiation, might be indirectly affected by UV-Exposure of the host plant shoots. Bacteria and fungi could also be pathogenic for both plants and animals, although beneficial microorganisms and plant pathogens had received more attention than animal pathogens with respect to climate changes (Van de Staaij *et al.*, 2001).

Plant disease management still relies on the use of fungicides although alternative methods have been under investigation for a long time. The persistence of a chemical on the plant surface depends on weather conditions, in particular precipitation and temperature. Changes in duration, frequency and intensity of rain events alter the efficacy of fungicides since they can quickly wash away. The temperature influences the degradation of pesticides, and alters the morphology and physiology of plants affecting their penetration, translocation and mode of action (Elad and Pertot, 2014). Pathogen life style and activity of biological control agents can be also influenced by environmental factors, although there is very little information on the impact of climate change on plant disease biological control. These are living organisms, and different climate conditions regularly affect their efficacy. For example, *Trichoderma harzianum* T-39 is more active against grey mould at higher temperature and lower relative humidity level (Elad *et al.*, 1993).

The impact of climate change on plant diseases requires more research. Although this topic has developed rapidly in the last few decades, there are still some gaps that need clarification. The prediction and the management of climate change effects on plant health are influenced by interactions with global change drivers. The models of plant disease development can vary under different forms of climate change, requiring different management strategies based on more participatory approaches and multidisciplinary science.

References

Agrios, G. N., 2005, Plant Pathology, Academic Press, San Diego.

Andresen, L. C., Dungait, J. A. J., Bol, R., Selsted, M. B., Ambus, P. and Michelsen, A., 2014, Bacteria and fungi respond differently to multifactorial climate change in a temperate heathland, traced with 13C-glycine and FACE CO_2. PLoS One, 9(1), e85070.

Angelotti, F., Scapin, C. R., Tessmann, D. J. Vida, J. B. and Canteri, M. G., 2014, The effect of temperature, leaf wetness and light on development of grapevine rust. Australas. Plant Pathol., 43: 9-13.

Anne legreve and Etienne Duveiller, 2010, Preventing potential disease and pest epidemics under changing climate. Climate change and crop production: 50-57.

Bergot, M., Cloppet, E., Perarnaud, V., Deque, M., Marcais, B. and Desprez-Loustau, M. L., 2004, Simulation of potential range expansion of oak disease caused by *Phytophthora cinnamomi* under climate change. Global Change Biology, 10: 1539-1552.

Booth, T. H., Jovanovic, T., Old, K. M. and Dudzinski, M. J., 2000, Climatic mapping to identify high-risk areas for *Cylindrocladium quinqueseptatum* leaf blight on eucalypts in mainland South East Asia and around the word. Environ. Pollution., 108: 365-372.

Bosch, J., Carrascal, L. M., Duran, L., Walker, S. and Fisher, M. C., 2007, Climate change and outbreaks of amphibian chytridiomycosis in a montane area of Central Spain. Proc. R. Soc., 274: 253-260.

Braga, M. R., Aidar, M. P. M., Marabesi, M. A. and Godoy, J. R. L., 2006, Effects of elevated CO_2 on the phytoalexin production of two soybean cultivars differing in the resistance to stem canker disease. Environ. and Experimental Botany, 58: 85-92.

Burdon, J. J., Thrall P. H. and Ericson, A. L., 2006. The current and future dynamics of disease in plant communities. Annual Rev. Phytopathol., 44: 19-39.

Canadell, J. G., Le Quire, C., Raupach, M. R., Field, C. B., Buitenhuis, E. T., Ciais, P., Conway, T.J., Gillett, N. P., Houghton, R. A. and Marland, G., 2007, Contributions to accelerating atmospheric CO_2 growth from economic activity, carbon intensity, and efficiency of natural sinks. Proceedings of the National Academy of Sci., 104: 18660-18870.

Chakraborty, S. and Datta, S., 2003, How will plant pathogens adapt to host plant resistance at elevated CO_2 under a changing climate. New Pathologist, 159: 733-742.

Chakraborty, S. and Pangga, I. B., 2004, Plant disease and climate change. In: Gillings, M.; Holmes, A. (Ed.) Plant microb., London: BIOS Scientific, 163-180.

Chakraborty, S., 2005, Potential impact of climate change on plant–pathogen interactions. Aust. Plant Pathol., 34: 443-448.

Chakraborty, S., 2013, Migrate or evolve: Options for plant pathogens under climate change. Glob. Chang. Biol., 19: 1985-2000.

Chakraborty, S., Pangga, I. B., Lupton, J., Hart, L., Room, P. M. and Yates, D., 2000, Production and dispersal of *Colletotrichum gloeosporioides* spores on *Stylosanthes scabra* under elevated CO_2. Environ. Pollut., 108: 381-387.

Chakraborty, S., Tiedemann, A.V. and Teng, P. S., 2004, Climate change: potential impact on plant diseases. Environ. Pollut., 108: 317-326.

Coakley, S. M., Scherm, H., and Chakraborty, S. 1999. Climate change and plant disease management. Annu. Rev. Phytopathol. 37: 399-426.

Dalla Pria, M., Christiano, R. C. S., Furtado, E. L., Amorim, L., and Bergamin Filho, A., 2006. Effect of temperature and leaf wetness duration on infection of sweet oranges by Asiatic citrus canker. Plant Pathol., 55: 657-663.

Djanaguiraman, M., Prasad, P. V. V., Seppanen, M., 2010, High night temperature decreases leaf photosynthesis and pollen function in grain sorghum. Plant Physiol. Biochem., 48: 999-1007.

Eastburn, D. M., Degennaro, M. M., Delucia, E. H., Dermody, O. and McElrone, A. J., 2010, Elevated atmospheric carbon dioxide and ozone alter soybean diseases. Global Change Biol., 16: 320-330.

Elad, Y. and Pertot, I., 2014, Climate change impacts on plant pathogens and plant diseases. J. Crop Improv., 28: 99-139.

Elad, Y., Zimand, G., Zaqs, Y., Zuriel, S. and Chet, I., 1993, Biological and integrated control of cucumber grey mould (*Botrytis cinerea*) under commercial greenhouse conditions. Plant Pathol., 42: 324-332.

Garrett, K. A., Nita, M., De Wolf, E. D., Gomez, L. and Sparks, A. H., 2009, Plant pathogens as indicators of climate change. In T. Letcher (Ed), Climate change: observed impacts on planet Earth, pp. 425-437.

Ghini, R. 2005, Mudanças climáticas globais e doenças de plantas. Jaguariúna: Embrapa Meio Ambiente, 104pp.

Gioria, R., Brunelli, K. R. and Kobori, R. F., 2008, Impacto potencial das mudanças climáticas sobre as doenças de hortaliças: tomate, um estudo de caso. Summa Phytopathologica, 34: 121-122.

IPCC 2007, Climate change, Synthesis Report. Contribution of Working Group I, II and III to the Fourth Assessment Report of the Intergovernmental Panel on Climate Change. Pachauri RK, Reisinger A, (Eds.) IPCC, Geneva, Switzerland.

IPCC. Climate change 2001: the scientific basis. Geneva: IPCC/WG I., 881 p. (Assessment Report, 3).

Karnosky, D. F. Percy, K. E., Xiang, B. X., Callan, B., Noormets, A., Mankovska, B., Hopkin, A., Sober, J., Jones, W., Dickson, R. E. and Isebrands, J. G., 2002, Interacting elevated CO_2 and tropospheric O_3 predisposes aspen (Populus tremuloides Michx.) to infection by rust (*Melampsora medusae* f. sp tremuloidae). Global Change Biology, 8: 329-338.

Kobayashi, T., Ishiguro, K., Nakajima, T., Kim, H.Y., Okada, M. And Kobayashi, K., 2006, Effects of elevated atmospheric CO_2 concentration on the infection of rice blast and sheath blight. Phytopathology, 96: 425-431.

Kudela, V., 2009, Potential impact of climate change on geographic distribution of plant pathogenic bacteria in Central Europe. Plant Protect. Sci., 45: 27-32.

Ladanyi, M. and Horvath, L., 2010, A review of the potential climate change impact on insect populations – general and agricultural aspects. Appl. Ecol. Environ. Res., 8: 143-152.

Lambers, H., Chapin, F. S., and Pons, T. L., 2008, Plant Physiological Ecology, 2nd edn. Springer Science Business Media LLC.

Laxman, S. R., Taru S., Puchakayala, M. and Parimal, S., 2017, Effect of temperature on growth and sporulation of rice leaf blast pathogen *Magnaporthe oryzae*. Int. J. Curr. Microbiol. App. Sci. 6(3): 394-401.

Legreve, A. and Duveiller, E., 2010, Preventing potential diseases and pest epidemics under a changing climate. In M. P. Reynolds (Ed.), *Climate change and crop production*: 50–70. Wallingford: CABI.

Lopes, C. A., Reis, A. and Shimoyama, N., 2008, Impacto potencial das mudanças climáticas sobre as doenças da batata no Brasil. In: Ghini, R.; Hamada, E. Mudanças climáticas: impactos sobre doenças de plantas no Brasil. Brasília: Embrapa/SCT, 3: 75-92.

Mamta Sharma, and Suresh Pande, 2016, Dry root rot (*Rhizoctonia bataticola* (Taub.) Butler): an emerging disease of chickpea – where do we stand. Archiv. of Phytopathol. and Plant Protec., 53: 1-16.

Mboup, M., Bahri, B., Leconte, M., De Vallavieille-Pope, C., Kaltz, O. and Enjalbert, J., 2012, Genetic structure and local adaptation of European wheat yellow rust populations: the role of temperature-specific adaptation. Evol. Appl. 5: 341–352.

Mcelrone, A. J., Hamilton, J. G., Krafnick, A. J., Aldea, M., Knepp, R. G., and De Lucia, E. H. 2010, Combined effects of elevated CO_2 and natural climatic variation on leaf spot diseases of red bud and sweetgum trees. Environ. Pollut., 158: 108-114.

Pangga, I. B., Chakraborty, S. C. and Yates, D., 2004, Canopy size and induced resistance in *Stylosanthes scabra* determine anthracnose severity at high CO_2. Phytopathol., 4: 221-227.

Rosenzweig, C., Iglesias, A., Yang, X. B., Epstein, P. R. and Chivian, E., 2001, Climate change and extreme weather events. Global Change & Human Health, 2: 90-104.

Savary, S., Nelson, A., Adam, H., Sparks, Willocquet, L., Duveiller, E. and Mahuku, G., 2011, International agricultural research tackling the effects of global and climate changes on plant diseases in the developing world, Plant Dis., 95: 1204-1216.

Siegenthaler, U., Stocker, T. F., Monnin, E., Luthi, D., Schwander, J., Stauffer, B., Raynaud, D., Barnola, J. M., Fischer, H., Masson-Delmotte, V. and Jouzel, J., 2005, Stable carbon cycle-climate relationship during the late pleistocene. Sci., 310: 1313-1317.

Spahni, R., Chappellaz, J., Stocker, T. J., Loulergue, L., Hausammann, G., Kawamura, K., Fluckiger, J., Schwander, J., Raynaud, D., Masson-Delmotte, V. and Jouzel, J., 2005, Atmospheric methane and nitrous oxide of the late pleistocene from Antarctic ice cores. Sci., 310: 1317-1321.

Stern, N., 2007, The Economics of Climate Change: The Stern Review. Cambridge University Press, Cambridge, UK.

Strange, R. N. and Scott, P. R., 2005, Plant disease: A threat to global food security. Ann. Rev. of Phytopathol., 43: 83-116.

Tiedemann, A. V. and Firsching, K. H., 2000, Interactive effects of elevated ozone and carbon dioxide on growth and yield of leaf rust-infected versus non-infected wheat. Environ. Pollut., 108: 357-363.

Van de Staaij, Rozema, J. W. M., Van Beem, J. A. and Aerts R., 2001, Increased solar UV-B radiation may reduce infection by arbuscular mycorrhizal fungi (AMF) in dune grassland plants: evidence from five years of field exposure. Plant Ecol., 154: 171–177

8

Impact of Climate Change on Vector Borne Diseases of Crop Plants

D.S. Aswathanarayana, Y.S.Amaresh, S.B. Gowdar and A.G. Sreenivas

Introduction

The information of how climate change is likely to influence future virus disease epidemics in cultivated plants and natural vegetation is of great importance to both global food security and natural ecosystems. However, obtaining such knowledge is hampered by the complex effects of climate alterations on the behavior of diverse types of vectors and the ease by which previously unknown viruses can emerge. Recently, there has been encouraging progress in understanding the likely effects of some climate change parameters, especially over the effects of elevated CO_2, temperature and rainfall-related parameters, upon a small number of important plant viruses and several key insect vectors, especially aphids (Jones, 2016).

Atmospheric greenhouse gas concentrations have reached levels that are unprecedented for 800,000 years. They continue to accumulate in the earth's atmosphere at an alarming rate causing climate change to accelerate with increasingly serious consequences for mankind and the planet's other life forms. Since the start of the industrial era in 1750, the global average concentration of carbon dioxide (CO_2) in the atmosphere has increased by 41 %, methane by 160 %, and nitrous oxide by 20 %. Atmospheric CO_2 levels are currently about 400 μmol-1 and elevated CO_2 levels are forecast to rise to 650 μmol-1 by the year 2100 (IPPC, 2014). The International Panel on Climate Changes' Fifth Assessment Report predicted that: (i) the global surface temperature increase by the end of the 21st century is likely to exceed 1.5°C relative to the 1850–1900 period for most scenarios and is likely to exceed 2.0°C for many scenarios; (ii) the global water cycle will change, with increases in disparity between wet and dry regions, as well as wet and dry seasons; (iii) the oceans will continue to warm, with heat extending to the deep ocean, affecting circulation patterns; (iv) decreases are very likely in Arctic sea ice cover, Northern Hemisphere spring snow cover, and global glacier volume; (v) global mean sea level will continue to rise at a rate very likely to exceed the rate of the past four decades; (vi) changes in climate will cause an increase in the rate of CO_2 production and its increased uptake by oceans will increase their acidification; (vii) future surface temperatures will be largely determined by cumulative CO_2, which means climate change will continue even if CO_2 emissions are stopped; and (viii) only substantial and sustained global emission reductions would help to reduce climate risks (IPPC, 2014).

Plant pests and pathogens must evolve or migrate to survive as per climate change. Accelerated evolution prompted by a changing environment drives development of variants potentially better adapted to the new conditions and changing geographic distributions bring together diverse lineages thereby increasing diversity (Chackraborty, 2013). Observations on changes in crop pests and pathogen distributions over the 20th century suggested growing agricultural production and trade have been the most important factors disseminating them, but global warming is also causing them to move poleward. Poleward shifts of >600 crop pests and pathogens averaged 27 kms/decade since the 1960s (Bebber *et al.*, 2013). Plant hosts, vectors and viruses are influenced by (i) the direct consequences of climate change, especially altered rainfall patterns, increased temperature, greenhouse gases, drought and greater wind speeds; and (ii) indirectly by things like regional alterations in areas cropped, ranges of crops grown, cultivation systems, distribution and abundance of vectors and weed or cultivated reservoir hosts. In turn, these factors influence geographic ranges and relative abundance of viruses, their rates of spread, the effectiveness of host resistances, the physiology of host–virus interactions, the rate of virus evolution and host adaptation and the effectiveness of control measures (Jones, 2009).

Jones and Barbetti (2012) discussed critical features of virus and vector life cycles that need to be considered in order to understand likely climate change scenarios. Some of the most important of these are:

- Viruses can be generalists that have broad host ranges but are often poorly adapted to individual hosts, or specialists which have narrow host ranges but are well adapted to their hosts. Generalists tend to be better adapted than specialists to alterations in host range under climate change scenarios.
- Due to their simple genomes, viruses tend to adapt quickly to environmental changes which explains by. Anderson *et al.* (2004) calculated that they cause nearly half (47%) of emerging infectious diseases of plants.
- Host plants and virus vectors react to climate change by alterations in their physiology, morphology and phenology.
- Cultivated host plants can be moved between different climatic zones by man's activities, but many non cultivated species face extinction because

of their inability to adapt to a changing climate fast enough or disperse their seeds or other propagules to new regions.

- Insects are the most important virus vectors and the most important types of insect vectors are aphids, whiteflies, and thrips, but other insects, such as leafhoppers, plant hoppers, mealybugs and beetles, also transmit some viruses. Mites are also important vectors.
- Aphid-borne viruses (*e.g.* potyviruses) are the most widespread and damaging viruses of cultivated plants in temperate regions, but whiteflyborne viruses (*e.g.* begomoviruses) and thrips-borne viruses (*e.g.* tospoviruses) are often the most important in regions with tropical and subtropical climates. All three insect vector types are important in protected cropping.
- Soil-borne viruses are transmitted from roots of infected plants to healthy plant roots by ectoparasitic nematodes or "fungus-like" organisms (oomycetes and protists) that are obligate root parasites. Soils infested with virus-infected vector resting spores can be dispersed by flooding activities, and some viruses are water-borne, being spreading over long distances by flooding or in irrigation and drainage channels.
- Contact-transmitted viruses are generally ones that reach high concentrations in the plant and have particles that are relatively stable outside the infected cell.

Climate change is a major environmental challenge worldwide. Green house gases (GHG) *viz.*, water vapour (H_2O), carbon dioxide (CO_2), Methane (CH_4), nitrous oxide (N_2O), hydrofluorocarbons (HFCs) and Ozone (O_3) in the atmosphere trap reflected radiation to warm the earth surface (Mahato, 2014). Human activities are widely involved in increasing global climate changes that directly influences the ecology (Ahanger *et al.*, 2013). This global climate change influences all the 3 major elements of disease triangle, *viz.*, host, pathogen and environment (Legreve and Duveiller, 2010).

The challenge that the increasing pace of global warming and climate instability pose to mankind's ability to manage pests and diseases of cultivated plants and natural vegetation is cause for great concern. This is because of the consequent increase in global food insecurity at a time of rapid population increase and the accompanying loss of natural plant biodiversity. Many reviews have been written about the anticipated influences of diverse climate change parameters on (i) fungal pathogens and the plant diseases they cause, and (ii) insect pest populations and the plant damage they induce. A small number of these reviews also provided examples of the likely effects of climate change parameters on plant viral pathogens or the roles of virus vectors (*e.g.* Garrett *et al.*, 2006; Luck *et al.*, 2011; Southerst *et al.*, 2011). However, only three reviews include the likely effects of climate change on plant virus diseases globally as a major focus (Canto *et al.*, 2009; Jones, 2009; Jones and Barbetti, 2012). Canto *et al.* (2009) discussed probable climate change effects on hosts and vectors that influence the spread of hemipteran-borne viruses, and Jones (2009) summarized likely climate change effects on plant virus epidemics as part of an article also covering virus origins, emergence, and evolution. Jones and Barbetti (2012) used climatic and biological frameworks to establish the probable effects of direct and indirect climate change parameters on the many vectors, virus, and host parameters that represent the full spectrum of plant virus pathosystems. As regards viruses, their review addressed the many multifaceted and geographically distinct ramifications of climate change likely to influence epidemics of vector-borne and convector-borne plant virus pathogens worldwide. Their approach provided comprehensive coverage of international research then available that illustrated the likely influences of climate change on virus diseases of cultivated plants and natural vegetation in different regions of the world. They concluded that (i) climate change is likely to modify many critical virus epidemic components in different ways often resulting in epidemic enhancement but sometimes having the opposite effect, depending on the pathosystem and circumstances; (ii) with vector-borne pathosystems and new encounter scenarios, the complication of having to consider effects of climate change parameters on diverse types of vectors and emergence of previously unknown virus pathogens added important additional variables; and (iii) the increasing difficulties in controlling damaging plant virus epidemics predicted to arise from future climate instability warrants considerable research effort to safeguard world food security and biodiversity.

The vector diversity is key to understand plant virus epidemiology. The plant disease dynamics may be more constrained by the host range of vectors than the host range of virus. In general plant viruses have a low diversity of vectors and a diversity of hosts and the largest proportion of viruses fall in to the category of host generalist/vector specialist (Reddy, 2010). A majority of vector- transmitted viruses (58.4 %) have a single known vector species, whereas only 9.9 per cent of such viruses have a single known host plant. In contrast, the proportion of host specialists among viruses without vectors (18.0 %) is significantly higher (Reddy, 2010).

Direct Climate Effects on Vector Borne Viruses

Greenhouse Gases

Carbon fertilization due to increased atmospheric CO_2 alters plant growth, morphology, biomass, physiology, metabolic pathways and microclimates. It increases the growth and yield of most crop species. This increase is mainly caused by increased rates of photosynthesis and/or increased water use efficiency. Efficiency of photosynthesis depends on intercellular CO_2 which is related to atmospheric CO_2 concentration. However, the potential for greater crop yields arising from increasing ambient CO_2 levels in temperate higher latitude regions, and cool mountainous regions in otherwise warmer areas, is unlikely to be realized fully. This is because the impact of CO_2 fertilization may be limited by deficiencies in other nutrients, increased respiratory C demand due to higher temperatures, or diminished water supply. eCO_2 influences secondary metabolite pathways, altering (i) the nutritious value of leaves to arthropod virus vectors, and (ii) the patterns of gene expression of defense signaling routes against both vectors and viruses. Alterations in plant growth or morphology influence vector behavior. These eCO_2 -induced changes all affect the way viruses and vectors interact with their plant hosts and influence the spatial and temporal dynamics of virus epidemics (Canto *et al.*, 2009; Jones and Barbetti, 2012).

Impact of Elevated CO_2 on Plant Viruses

Canto *et al.* (2009) and Jones and Barbetti (2012) found little information on the direct effects of eCO_2 on virus infection of plants, except that it decreased stunting symptoms in BYDV-infected oat plants but increased their biomass and water use efficiency (Malmstrom and Field, 1997), increased Potato virus Y (PVY) resistance in tobacco plants (Matros *et al.*, 2006), and had the potential to suppress pathogen-induced virus resistance (Garrett *et al.*, 2006). Zhang *et al.* (2015) found that eCO_2 lowered disease severity and modified plant defense responses in Tobacco mosaic virus (TMV)-infected tomato plants. This increased virus resistance and lowered disease incidence. Huang *et al.* (2012) also found that eCO_2 lowered disease severity and increased above ground plant height and biomass in Tomato yellow leaf curl virus (TYLCV)-infected tomato plants growing in open-top chambers in the field.

Impact of Elevated CO_2 on Arthropod Vectors in General

eCO_2 alters insect herbivory by altering both the defense chemistry and signaling of plants and their nutritional and water contents, but the responses of herbivorous insects to these alterations are highly variable. It increases the production of salicylic acid but suppresses the production of jasmonates and ethylene, and these differential responses affect secondary metabolite pathways (Zavala *et al.*, 2013). The indirect effects of eCO_2 on insect herbivores, such as aphid, thrips and whitefly virus vectors, include changes in their feeding, growth rates, fecundity and population density. These changes are mostly mediated by its effects on host plant quality including changes in host morphology, diversity, abundance, biochemistry, physiology, and composition such as increased C:N ratios and altered concentrations of soluble and nonstructural carbohydrates, starch and soluble proteins (Cornelissen, 2011; Finlay and Luck, 2011). There may also be direct effects of eCO_2 on insect physiology, behavior, and life history parameters (Cornelissen, 2011). Canto *et al.* (2009) and Jones and Barbetti (2012) described the information on the effects of eCO_2 on insect vectors of plant viruses. In brief, eCO_2 influenced different insect vector species in different ways. It had mixed effects on aphids which were highly species specific. With whitefly vector species, it had no effect on *Bemisia tabaci* (sweet potato whitefly) but affected *Trialeurodes vaporariorum* (glasshouse whitefly) negatively. With thrips vectors, it had little effect on populations of *Frankliniella occidentalis* (western flower thrips), but increased the population sizes of some other thrips vector species and may compromise host resistance to some insect vectors.

Impact of Elevated CO_2 on Aphid Vectors

Aphids are soft bodied homopteran insects which are one of the most serious agricultural insect pests as they cause major economic lossess in several crops world wide. Aphids are most common vectors of plant viruses and aphid borne non persistently transmitted virus diseases are of great economic importance in several annual cropping systems (Tomlinson, 1987). Aphid borne viruses which infect many economically important crops in India are banana bract mosaic virus, banana bunchy top virus, bean common mosaic virus, bean yellow mosaic virus cardamom mosaic virus, chilli veinal mottle virus, cowpea aphid borne virus, cucumber mosaic virus and zucchini yellow mosaic virus (Reddy, 2010).

Oehme *et al.* (2013) examined the effects of eCO_2 on the performance of the important vector species *Rhopalosiphum padi* (oat aphid) and *Myzus persicae* (green peach aphid) on plants of wheat and canola (oilseed rape), respectively. The relative growth rate of *R. padi* increased in wheat but that of *M. persicae* decreased in canola. An increase in the concentrations of the carbohydrates fructose and glucose occurred in the phloem sap of wheat but not of canola, but whether this or changes in concentrations of certain amino acid were responsible for these two aphid species–host combinations was not established. Dader *et al.* (2016) studied the life history and feeding behavior of *M. persicae* on pepper plants under ambient CO_2 and eCO_2 conditions, and the effect of these conditions on plant growth and leaf chemistry. Pepper plants fixed less nitrogen but were taller with greater biomass and canopy temperature under eCO_2. eCO_2 conditions increased the pre reproductive period of *M. persicae* and decreased its fecundity and salivation into sieve tubes, but did not alter phloem ingestion.

Impact of Elevated CO_2 on Other Arthropod Vectors

There are few recent examples of studies on the effects of eCO_2 on insect vectors other than aphids. With *B. tabaci* biotype B (¼Middle East-Asia Minor 1), an important virus vector biotype, oviposition, nymphal survival, and reproduction were not significantly affected by eCO_2 conditions (Cumutte *et al.*, 2014). When Wang *et al.* (2014) used open-top chambers to examine the effects of eCO_2 on the interaction between *B. tabaci* and its parasitoid *Encarsia formosa* on plants of Bt cotton and nontransgenic cotton, significantly longer egg-adult development times and greater nymph mortality occurred on both types of cotton. However, transgenic Bt cotton made no difference to the development, survivorship lifespan or fecundity of either insect and there were no significant effects on parasitism by *E. formosa.*

Impact of Elevated CO_2 on Nematode Vectors

The population density of the nematode virus vector *Longidorus elongatus* in grass-dominated pasture was increased by eCO_2 (Yeates and Newton, 2009; Yeates *et al.*, 2003). However, sometimes the population densities of root-feeding Longidoridae in soil do not benefit from the increased root biomass that occurs with eCO_2. This may be due to factors such as enhancement of plant defense mechanisms against root-feeding nematodes or diminished availability of soil nitrogen under eCO_2 conditions (Cesarz *et al.*, 2015). There still appears to be no published information on the influences of eCO_2 on fungal virus vectors.

Temperature

Viruses have different temperature optima for multiplication within their host plants, some being adapted to warmer regions and others to cooler ones. Such optima are highest in viruses adapted to invade hosts growing in low- land tropical environments and lowest in viruses adapted to cold climates, such as those in cool temperate zones or at high altitude in mountainous regions. Different species of host plants and virus vectors (insects, mites, nematodes, fungi) also have diverse temperature optima under which they flourish, some being adapted to warmer climates and others to cooler ones. As the mean temperature increases, generalist viruses with high-temperature optima for multiplication within their host plants that are adapted to warmer regions and infect many different host species are likely to expand their geographical ranges from the areas with tropical or subtropical climates they currently occupy. For vector-borne viruses, such expansion would be limited if the ranges of their vectors were to remain unchanged, but the opposite scenario is predicted for key tropical vectors of such viruses, such as *B. tabaci* and *T. palmi* (melon thrips). Warmer winters can increase virus epidemics in regions with temperate climates by increasing overwintering virus and vector survival. By contrast, hotter, drier summers decrease virus epidemics in rainfed crops in regions with Mediterranean climates by decreasing virus and vector over-summering. Unusually high air temperatures can reduce insect vector populations so prolonged heat waves may diminish epidemics of the viruses they transmit.

Impact of Temperature on Viruses

Increasing temperature alters host plant physiology, metabolic pathways, nutritional status, morphology, and phenology (Canto *et al.*, 2009; Jones and Barbetti, 2012). Rising mean temperature and heat stress increase general plant susceptibility to virus infection and decrease the effectiveness of temperature-sensitive single-gene resistances. Increased mean temperature also alters rates of virus multiplication, systemic movement and seed transmission and influences the multiplication and systemic movement of individual viruses present in mixed infection. Moreover, it modifies virus evolution rates and selection pressures which can lead to development of virulent virus strains with broadened natural host ranges, higher virus multiplication rates in reservoir hosts, and increased vector transmission efficiencies. Prolonged heat waves are likely to cause remission of virus symptoms in infected plants by reducing their virus contents. In some instances, prolonged heat waves might eliminate virus infections from growing plants where systemic invasion is already incomplete or the virus involved has unstable particles.

Jones and Barbetti, 2012; Jones (2014) described the likely effects of elevated temperature on potato virus epidemics worldwide. Potato-infecting viruses best adapted to warmer regions (*e.g.* Potato leaf roll virus (PLRV) and Potato yellow vein virus (PYVV)) are likely to expand from areas they occupy now to areas of higher latitude previously too cool for them, and to formerly cooler higher elevations in mountainous regions within the tropics or subtropics. A similar scenario is likely with the soil-borne viruses of wheat which include furoviruses, *e.g.* Soil-borne wheat mosaic virus (SbWMV), and bymoviruses, *e.g.* Wheat spindle streak mosaic virus (WSSV). Such wheat viruses are adapted to temperate conditions and mostly occur in cooler parts of Europe, Asia, North America and New Zealand (Cox *et al.*, 2014). When the soil moisture is sufficient for activity of the motile zoospores of their vector, the former viruses are likely to contract to areas of higher latitude or higher elevations in mountainous regions within the tropics and subtropics including areas previously too cold for the wheat crop, while the latter viruses are likely to expand to areas of higher latitude previously too cool for them and to formerly cooler higher elevations in mountainous regions within the tropics or subtropics (Jones, 2014a).

Impact of Temperature on Aphid Vectors

Aphids react strongly to small changes in mean temperatures due to their mobility, low developmental threshold temperatures, short generation times, high capacity for reproduction, and ability to make rapid life history and behavioral changes. An additional five generations of aphids/year are predicted in temperate zones from a warming of 2°C. Thus the risk of serious epidemics of aphid-transmitted viruses increases as their populations and activities increase. In temperate regions, survival of aphid vectors is expected to increase with milder mean winter temperatures and higher mean summer temperatures are likely to increase their development and reproductive rates.

Chen *et al.* (2013) examined the effect of elevated temperature on the performance of the vector species *Aphis craccivora* (cowpea aphid). The aphids were reared at four temperatures: current midsummer mean of 28.6°C in subtropical Taiwan, +1.4°°C, +3.6°C, and +6.4°C. The aphids experienced either constant or daily oscillating (from - 2.9°C to +3.6°C) temperatures. As temperatures increased, so did negative effects on life history traits and demographic parameters. Also, compared with those reared at constant temperatures, aphids reared under oscillating temperatures developed more slowly and had longer mean generation times, but their net reproductive rate was higher.

Temperature influences the activity, abundance and distribution of parasitoids that suppress aphid vector populations and this knowledge is important in relation to aphid transmission of viruses under projected increased temperature global warming scenarios

Impact of Temperature on Whitefly Vectors

As with aphid vectors, whitefly vectors also react strongly to climatic changes due to their short generation times, high reproductive capacity and ability to make rapid life history and behavioral changes. *B. tabaci* and *T. vaporariorum* transmit different whitefly-transmitted viruses. Both flourish under warm conditions but *B. tabaci* is less cold tolerant than *T. vaporariorum*. With *B. tabaci*, 25-28°°C is optimal for development, and much shorter adult to-adult generation times occur at high (31-33°C) than low (17°C) temperatures. For *B. tabaci* to flourish, an average monthly temperature of atleast 21°C in the hottest month of the year, and a dry season with a period of 4 months of monthly rainfall of less than 80 mm, Because of increasing mean winter temperatures in places formally too cold for it in winter, *B. tabaci* is tending to displace *T. vaporariorum* which is increasing its distribution in formerly cooler regions. In turn, this shift in vector distribution is influencing whitefly-transmitted virus distributions in different parts of the world, and damaging epidemics of *B. tabaci*-transmitted begomoviruses are becoming more widespread (Jones and Barbetti, 2012).

The whitefly, *Bemisia tabaci* (Gennadius) (Homoptera: Aleyrodidae) is an important agricultural pest in tropical and subtropical regions and has colonized on the plants in green houses of semi-temperate countries. Importance of *B. tabaci* has increased in globally since the 1990's as a serious pest of vegetable, fibre and root crops (Varma and Malathi, 2003). Geminiviruses transmitted by the whitefly *B. tabaci* are widespread in tropical and subtropical regions of the world, where they cause numerous diseases in dicotyledonous plants including cassava, pulses, vegetables, tobacco and cotton (Muniyappa and Veeresh, 1984; Harrison, 1985; Brown and Bird, 1992).

B biotype of *B. tabaci* has been first recorded in 1999 in Kolar district of Karnataka, India, during summer (March- June) season. The economic importance of B biotype becomes immediately evident when number of whiteflies on tomato, cabbage, and other vegetable crops. The most significant damage was due to an epidemic of tomato leaf curl virus disease, which has since continued to affect tomato production seriously in India (Banks *et al.*, 2001). Currently B biotype is present in most parts of South India, and has spread from the original point of introduction in Kolar district since 1999. The B. biotype has spreaded about 600 km (Mysore to Raichur) in north-south and about 250 km (Shimogga to Chittor) in the East-West directions (Rekha *et al.*, 2005).

In India also whitefly transmitted begomoviruses in recent years are appearing in moderate to severe form and whitefly (*B. tabaci*) can serves as potential vector (Muniyappa and Veeresh, 1984) for the following emerging diseases *viz.*, Tomato leaf curl virus disease (Muniyappa *et al.*, 2000 and 2003) Indian Cassava Mosaic Virus (Maruthi *et al.*, 2002), Jatropha Mosaic Virus (Aswathanarayana *et al.*, 2005 and 2006), Hibiscuss Leaf Curl Virus (Rajeshwari *et al.*, 2005), Sunflower Leaf Curl Virus (Govindappa *et al.*, 2011), Mesta Leaf Curl Virus (Humma Ambuja, 2015), Mungbean Yellow Mosaic Virus (Deepa *et al.*, 2017; Meti *et al.*, 2017). Among the various biotypes of *B. tabaci* prevailing in India, B biotype can adapt well in change in climate and have more efficient in transmission of Gemini viruses (Shankarappa *et al.*, 2006).

Guo *et al.* (2013) compared the survival, development and reproduction of the biotype B of *B. tabaci* over five generations at 27°C, 31°C, and 35°C. At 27°C, its survival, development and fecundity were stable, but its fecundity declined in the fourth and fifth generations at 31°C. At 35°C, egg hatching, immature survival rate, and fecundity all declined in the fourth and fifth generations. When held at 37°C for one generation fecundity was < 10 %. It therefore has the potential to continue acting as a virus vector under extreme elevated temperature climatic scenarios. Cumutte *et al.* (2014) studied the effects of elevated temperature on life history parameters of *B. tabaci* biotype B. Its oviposition, nymphal survival, and reproduction were all affected negatively, with net reproductive success declining to 36 per cent at 33°C. Overall, 28°C was most favorable for its fitness and 28-33°C for its reproduction, and temperature effects on the host plant had minimal effects on whitefly reproduction.

Impact of Temperature on Other Arthropod Vectors

Thrips, small soft bodied fringe winged insects belongs to order Thysanoptera, are known as pests of food and horticultural crops and acts as vectors of some plant diseases (David and Anantakrishnan, 2004). In India, *Thrips tabaci* and *T. Palmi* are important in terms of pest and vectors of tospoviruses. Successful interaction of thrips with their host plants appears to depend on a complex set of environmental, visual, tactile and chemical factors that appreciably influence their behaviour and physiology (David and Ananthakrishnan, 2004; Reddy, 2010). Although temperature is the main factor affecting onion thrips development, it is an open question of wheather the thermal constants are truely constant in different climatic conditions.

Increased mean temperatures can increase thrips vector populations by hastening their development rates leading to more generations per year. As different thrips species have different thermal thresholds, global warming is likely to change their species composition. For example, *T. palmi* is expected to expand into areas formerly too cold for it, displacing vector thrips species adapted to cooler temperatures. This is projected to cause expansion into new regions of several damaging tospoviruses it transmits. Similar considerations apply to viruses transmitted by leafhoppers, mealybugs, and eriophyid mites (Canto *et al.*, 2009; Jones, 2009; Jones and Barbetti, 2012). Li *et al.* (2014) exposed newly emerged adult *F. occidentalis* to 21°C, 26°C, and 33°C conditions. As temperature increased, their survival, longevity and reproduction diminished, but expression of four heat-shock proteins was upregulated.

Rainfall

Shifts in precipitation have significant impacts on vegetation patterns with attendant consequences for agriculture. Weather extremes result in excessive rainfall and flooding as well as severe drought (Anayamba *et al.*, 2014). Such weather extremes can create exceptional conditions for extensive virus disease outbreaks in cultivated and natural vegetation.

Impact of Rainfall on Arthropod-Transmitted Viruses and their Vectors

Epidemics of arthropod-transmitted viruses are projected to alter considerably as climate change alters the world's rainfall patterns, drought and flood events become more extreme, and wet seasons become increasingly less predictable in many world regions. For example, in mid-latitude or subtropical regions of Asia, Australia, Latin America, and Africa, the likelihood that damaging epidemics of *B. tabaci*-transmitted viruses, such as many begomoviruses, will occur increases if they develop the longer dry seasons between cropping periods that favor *B. tabaci* population increase. In contrast, in mid-latitude regions with Mediterranean-type climates and rainfed, winter cropping, hotter, dryer summer conditions may decrease epidemics of aphid-borne viruses by diminishing their abilities and those of their aphid vectors to over-summer. The increased frequency of heavy rainfall, extreme drought, or flood events expected to arise from climate change is likely to influence plant virus epidemics in different ways. An increasing frequency of heavy rainfall events is likely to slow them by washing insect vectors like aphids, thrips, and whiteflies off foliage, thereby diminishing vector populations, as occurs with monsoonal rains and *T. palmi* in northern Australia. Waterlogging due to flooding kills pupal life stages of thrips vectors in the soil and oviposition of *B. tabaci* is impaired by heavy rain, so slowing epidemics of the viruses they transmit by reducing insect vector populations. Flooding within annual crop growing seasons enhances the subsequent growth of weed and volunteer crop plant reservoirs of viruses and arthropod vectors, and its occurrence outside growing seasons increases subsequent growth of such reservoirs. Both scenarios would tend to enhance virus epidemics by increasing the arrivals of viruliferous arthropod vectors in crops and native vegetation. Conversely, prolonged drought stress on crop plants can limit arthropod vector population size and also restrict the size of weed and volunteer crop reservoirs of viruses and vectors. Both of these scenarios are likely to decrease virus epidemics by diminishing the arrival of viruliferous arthropod vectors in crops or native vegetation. However, in pot experiments virus infection delayed the onset of drought symptoms in plants under dry conditions which prolonged their lives so might possibly be beneficial in some cropping situations (Jones, 2009; Jones and Barbetti, 2012).

Impact of Rainfall on Contact-Transmitted and Water-Borne Viruses and Viroids

With epidemics of contact-transmitted viruses, greater frequency and intensity of summer storms associated with heavy rainfall or hail that increase rain splash and wounding of plants and also provide moisture, would be expected to accelerate their spread resulting in more severe virus epidemics. In addition, both drought and flood can influence survival of unprotected virus particles in contaminated soil and their abilities to act as sources of inoculum for spread to susceptible crops. For example, the dry, compacted or waterlogged soils typical of drought or flooding, respectively, favor survival of the stable, contact-transmitted virus TMV in plant debris in soil but well-aerated soils favor its inactivation (Jones and Barbetti, 2012). Mackie *et al.* (2016) described the widespread flooding caused by a rare extreme summer rainfall event in Dec. 2010 in a remote subtropical cropping region of coastal Western Australia. Plant material infected with the contact-transmitted Potato spindle tuber viroid (PSTVd) was moved to new areas by flood water allowing volunteer plants of tomato and pepper, both of which are PSTVd hosts, to flourish in dry areas never cultivated previously that were several kilometers away from any cropping areas. Also, the moisture in soils present after the flood abated permitted widespread, profuse growth of weed and native plant species.

Relative Humidity and Leaf Microclimates

Relative humidity determines things like duration of leaf wetness and is closely tied to temperature so is more predictable than many other climate parameters. High relative humidity is closely linked to local microclimates typified by moist leaf surfaces, whereas low humidity tends to cause a dry microclimate at the leaf surface. Relative humidity and leaf surface microclimates can have marked effects on the rate of population growth, behavior, survival, and movement of arthropod virus vectors. Also, high humidity favors development of entomopathogenic fungi that kill aphid vectors. Spread of contact-transmitted viruses in crops and pastures is favored by the lush plant growth and soft tender leaves that develop under conditions of high relative humidity. Such plants are much more susceptible to infection than the hard-leafed plants typical of low humidity conditions. This is because wounds develop more readily when growth is soft and viruses have to penetrate a plant's protective cuticle through wounds before they can invade damaged cells (Jones and Barbetti, 2012).

Wind Speed and Direction

Arthropod-borne viruses can be distributed over long distances by strong winds. For example, viruliferous aphids are sometimes transported over long distances initiating new virus infection foci far away from the initial virus inoculum source. Projected periodic increases in wind velocity associated with climate change are likely to increase dispersal of insect and mite vectors via wind currents, especially those vector species that already disperse readily. Prevailing winds are very important in determining the direction of spread of arthropod-transmitted viruses, downwind spread from a virus infection source having been demonstrated with different types of insect or mite vectors. Such winds are particularly important for spread of viruses with tiny, wingless mite vectors. Jones and Barbetti (2012) demonstrated wind-mediated transmission of Rice yellow mottle virus (RYMV) in irrigated rice crops (Sarra *et al.*, 2004).

Congdon *et al.* (2016) investigated wind-mediated contact transmission of Pea seed-borne mosaic virus (PSbMV) between pea plants in glasshouse experiments. PSbMV was transmitted when intertwining healthy and PSbMV infected plants were blown by an oscillating fan to simulate wind, but no passive contact transmission occurred. This suggested that by expanding initial seed-borne infection foci before aphid vector arrival, contact transmission might play an important role in PSbMV spread early in the life of a pea crop. Winds can affect the host-finding behavior of aphid vectors at various stages including takeoff, long-range migration, landing, and settling. Barton (2014) studied the effects of wind speed on the vector aphid *Aphis glycines* (soybean aphid) and its predator *H. axyridis*. In the field experiment in which plots were infested with both aphid and predator, aphids were more abundant at high wind pressure due to increased predation at lower wind strengths. This was because plant movement due to wind interfered with feeding by the predator and so decreased their predation.

Alterations in Weed or Cultivated Plant Reservoir Hosts

When seeds of weed species not previously found in a region are introduced as contaminants during trade in plants and plant products, seed-borne viruses can be introduced. Weed species formerly unable to establish because they possess specific temperature or rainfall requirements may subsequently be able to establish due to altered conditions resulting from climate change. In addition, the introduced weeds might provide favorable hosts for viruses or arthropod vectors already present, and the virus concerned might be damaging to an important crop grown locally. If weed control is inadequate, their presence would aggravate damaging epidemics of already occurring viruses in the vulnerable local crop. Moreover, cultivated plants introduced to regions formerly unsuited to them may be tolerant of viruses damaging to vulnerable cultivated plant species growing nearby. Their presence as major virus reservoirs would then enhance virus spread to, and virus epidemics within, the vulnerable crop which might be one grown traditionally in the region before it became suitable for plantings of the newly introduced crop. The increased vector population moving from the introduced cultivated plant reservoir would then drive an acceleration of virus epidemics in the vulnerable traditional crop (Jones, 2009; Jones and Barbetti, 2012).

Conclusion

In recent years intensive human interfere activities like urbanization, industrial development, reducing the cultivable agricultural lands and forest area resulted in the habitat change, affecting food chain, high input agriculture, and food insecurity to the growing population. It has great impact on the climate change with respect increase in temperature, precipitation and distribution pattern, elevation of green house gasses, wind and relative humidity. Ultimately the climate change has influence on cropping pattern, vector population (Thrips, whitefly, aphids, mite and leaf hoppers) which are potentially involve in transmission of plant viruses and alter the virus disease epidemics. Introduction of weeds and pests from outside the country can adjust to change in climate and establish well. The weeds an serve as reservoir for plant viruses and insects, which can enhance the epidemics of vector borne plant viruses. Finally by knowing the long term weather data of a geographical region and distribution pattern of insect vector from surveillance data can plan for development of forecasting models for establishment of insect vectors. With this Scientists can develop the effective management strategies for management of insect vector and vector borne diseases of plants.

References

Ahanger, R. A., Bhat, H. A., Bhat, T. A., Ganie, S. A., Lone, A. A., Wani, I. A., Ganai, S. A., Haq, S., Khan, O. A., Junaid, J. M. and Bhat, T. A., 2013. Impact of Climate Change on Plant Diseases. International Journal of Modern Plant and Animal Sciences, 1(3): 105- 115.

Anayamba, A., Small, J.L., Britch, S.C., Tucker, C.J., Pak, E.W., Reynolds, C.A., Crutchfield, J. and Linthicum, K.J., 2014. Recent weather extremes and impacts on agricultural production and vector-borne disease outbreak patterns. PLoS One 9 (3), e92538.

Anderson, P.K., Cunningham, A.A., Patel, N.G., Morales, F.J., Epstein, P.R. and Daszak, P., 2004, Emerging infectious diseases of plants: pathogen pollution, climate change and agrotechnology drivers. Trends Ecology and Evolution 19, 535–544.

Aswathanarayana, D. S., Shankarappa, K. S., Govindappa, M. R., Prameela, H. A., Gururaj Rao, M. R. and Rangaswamy, K. T., 2006, Natural occurrence of Jatropha Mosaic Virus Disease in India, Current Science, 91 (5): 584-586.

Aswathanarayana, D.S., Shankarappa, K.S., Prameela, H.A., Raghavendra, N., Keshava Murthy, K.V. and Rangaswamy, K.T., 2005, New Hosts of Begomoviruses in Karnataka. In: Second global conference: Plant health, Global wealth, pp. 14, Udaipur, 25-29 November, 2005.

Banks, G.K., Colvin, J., Chowda Reddy, R.V., Maruthi, M.N., Muniyappa, V., Venkatesh, H.M., Kiran Kumar, M., Padmaja, A.S., Beitia, F.J. and Seal, S.E., 2001, First report of the Bemisia tabaci B biotype in India and an associated tomato leaf curl virus disease epidemic. Plant Disease. 85: 231.

Barton, B.T., 2014, Reduced wind strengthens top-down control of an insect herbivore. Ecology, 95, 2375–2381.

Bebber, D.P., Ramotowski, M.A.T. and Gurr, S.J., 2013, Crop pests and pathogens move polewards in a warming world. National Climate Change, 3, 985–988.

Brown, J. K. and Bird, J., 1992, Whitefly transmitted diseases in America and Caribbean Basin: past and present. Plant Disease, 76: 220-225.

Canto, T., Aranda, M.A. and Fereres, A., 2009, Climate change effects on physiology and population processes of hosts and vectors that influence the spread of hemipteran-borne plant viruses. Glob. Chang. Biol. 15, 1884–1894.

Cesarz, S., Reich, P.B., Scheu, S., Ruess, L., Schaefer, M. and Eisenhauer, N., 2015, Nematode functional guilds, not trophic groups, reflect shifts in soil food webs and processes in response to interacting global change factors. Pedologia, 58, 23–32.

Chackraborty, S., 2013, Migrate or evolve: options for plant pathogens under climate change. Glob. Chang. Biol. 19, 1985–2000.

Chen, C.Y., Chiu, M.-C. and Kuo, M. H., 2013, Effect of warming with temperature oscillations on a low-latitude aphid, *Aphis craccivora*. Bull. Entomol. Res. 103, 406–413.

Congdon, B.S., Coutts, B.A., Renton, M. and Jones, R.A.C., 2016, Pea seed-borne mosaic virus: stability and wind-mediated contact transmission in field pea. Plant Dis. in press.

Cornelissen, T., 2011, Climate change and its effects on terrestrial insects and herbivory patterns. Neotrop. Entomol. 40, 155–163.

Cox, B.A., Luo, H. and Jones, R.A.C., 2014, *Polymyxa graminis* isolates from Australia: identi- fication in wheat roots and soil, molecular characterization and wide genetic diversity. Plant Dis. 98, 1567–1575.

Cumutte, L.V., Simmons, A.M. and Abd-Rabbou, S., 2014, Climate change and Bemisia tabaci (Hemiptera: Aleyrodidae): impacts of temperature and carbon dioxide on life history. Ann. Entomol. Soc. Am. 107, 933–943.

Dader, B., Fereres, A., Moreno, A. and Trebicki, P., 2016, Elevated CO_2 impacts bell pepper growth with consequences to *Myzus persicae* life history, feeding behaviour and virus transmission ability. Sci. Rep. 6, 19120.

David, B. V. and Ananthakrishnan, 2004, General and applied entomology, Tata McgrawHill Publication, New Delhi, pp. 1148.

Deepa, H., Govindappa, M.R., Kenganal, M., Kulkarni, S.A. and Biradar, S.A., 2017, Screening of Greengram Genotypes against Mungbean Yellow Mosaic Virus Diseases under Field Condition. Int. J. Pure App. Biosci., 5(2): 1049-1056.

Finlay, K.J. and Luck, J.E., 2011, Response of the bird cherry-oat aphid (Rhopalosiphum padi) to climate change in relation to its pest status, vectoring potential and function in a cropvector-virus pathosystem. Agric. Ecosyst. Environ. 144, 405–421.

Garrett, K.A., Dendy, S.P., Frank, E.E., Rouse, M.N. and Travers, S.E., 2006, Climate change effects on plant disease: genomes to ecosystems. Annu. Rev. Phytopathol. 44, 489–509.

Govindappa, M. R., Shankergoud, I., Shankarappa, K. S., Wickramaarachchi, W. A. R. T., Anjaneyareddy and Rangaswamy K. T., 2011, Molecular detection and partial characterization of Begomovirus associated with leaf curl disease of Sunflower (*Helianthus annuus*) in Southern India. Plant Pathology Journal, 1: 1-7.

Guo, J.-Y., Cong, L. and Wan, F.-H., 2013, Multiple generation effects of high temperature onthe development and fecundity of *Bemisia tabaci* (Gennadius) (Hemiptera: Aleyrodidae) biotype B. Insect Sci., 20, 541–549.

IPPC, 2014, Intergovernmental Panel on Climate Change, Fifth Assessment Report. http://www.ipcc.ch/activities/activities.shtml.

Harrison, B. D., 1985, Advances in geminivirus Research. Ann. Rew. Phytopathology. 23: 55-82.

Huang, L., Ren, Q., Sun, Y., Ye, L., Cao, H. and Ge, F., 2012, Lower incidence and severity of tomato virus in elevated CO_2 is accompanied by modulated plant induced defence in tomato. Plant Biol., 14, 905–913.

Humma Ambuja, 2015, Studies on leaf curl disease of mesta (*Hibiscus sabdariffa* L.). M.Sc. (Agri.) thesis submitted to University of Agricultural Sciences, Raichur, Karnataka, India. 119pp.

Jones, R.A.C., 2009, Plant virus emergence and evolution: origins, new encounter scenarios, factors driving emergence, effects of changing world conditions, and prospects for control. Virus Res., 141, 113–130.

Jones, R.A.C., 2016, Future scenarios for plant virus pathogens as climate change progresses. Adv Virus Res. 2016; 95: 87-147.

Jones, R.A.C. and Barbetti, M.J., 2012, Influence of climate change on plant disease infections and epidemics caused by viruses and bacteria. CAB Rev. 7 (22), 1–32.

Jones, R.A.C., 2014, Virus disease problems facing potato industries worldwide: viruses found, climate change implications, rationalising virus strain nomenclature and addressing the Potato virus Y issue. In: Navarre, R., Pavek, M.J. (Eds.), The Potato: Botany, Production and Uses. CABI, Wallingford, UK, pp. 202–224.

Legreve, A. and Duveiller, E., 2010. Preventing potential diseases and pest epidemics under a changing climate. p. 50-70. In M. P. Reynolds (ed.), Climate change and crop production. Wallingford, CABI.

Li, H.B., Zheng, Y.T., Sun, D.D., Wang, J.J. and Du, Y.Z., 2014, Combined effects of temperature and avermectins on life history and stress response of the western flower thrips, *Frankliniella occidentalis*. Pestic. Biochem. Physiol. 108, 42–48.

Luck, J., Spackman, M., Freeman, A., Trebicki, P., Griffiths, W., Finlay, K. and Chakraborty, S., 2011, Climate change and diseases of food crops. Plant Pathol., 60, 113–121.

Mackie, A.E., Rodoni, B.C., Barbetti, M.A., McKirdy, S.J. and Jones, R.A.C., 2016, Potato spindle tuber viroid: alternative host reservoirs and strain found in an isolated sub-tropical irrigation area. Eur. J. Plant Pathol. in press.

Mahato, A., 2014, Climate Change and its Impact on Agriculture. International Journal of Scientific and Research Publications, 4(4).www.ijsrp.org.

Malmstrom, C.M. and Field, C.B., 1997, Virus-induced differences in the response of oat plantsto elevated carbon dioxide. Plant Cell Environ., 20, 178–188.

Matros, A., Amme, S., Kettig, B., Buck-Sorlin, G.H., Sonnewald, U. and Mock, H.P., 2006, concentrations leads to modified profiles of secondary metabolites in tobacco cv. Samsun and to increased resistance against infection with Potato virus Y. Plant Cell Environ., 29, 126–137.

Maruthi, M.N., Colvin, J., Seal, S.E., Gibson, G. and Cooper, J., 2002, Co-adaptation between cassava mosaic geminiviruses and their local vector populations. Virus Res., 86: 71-85.

Meti Meghashree, Kenganal Mallikarjuna, Sunkad Gururaj, Aswathanarayana, D. S. and Shanwad, U. K., 2017, Spatial variability of mungbean Yellow mosaic virus (MYMV) in North Eastern Karnataka. Internat. J. Plant Protec., 10 (2): 420-428.

Muniyappa, V and Veeresh, G. K., 1984, Plant virus diseases transmitted by whitelies in Karnataka. Indian Acad. Sci. (Anim. Sci.), 93: 397-406.

Muniyappa, V., Maruthi, M.N., Babitha, C.R., Colvin, J., Briddon, R.W., and Rangaswamy, K.T., 2003, Characterisation of pumpkin yellow vein mosaic virus from India. Ann. Appl. Biol., 142, 323–331.

Muniyappa, V., Venkatesh, H.M., Ramappa, H.K., Kulkarni, R.S., Zeidan, M., Tarba, C.Y., Ghanim, M. and Czosnek, H., 2000, Tomato leaf curl virus from Bangalore (ToLCV-Ban4): sequence comparison with Indian ToLCV isolates, detection in plants and insects, and vector relationships. Arch. Virol., 145, 1583–1598.

Oehme, V., Hogy, P., Zebitz, C.P.W. and Fangmeier, A., 2013, Effects of elevated CO_2 concentrations on phloem concentrations of spring crops and aphid performance. J. Plant Interact., 8 (1), 74–84.

Rajeshwari, R., Reddy, R.V. C., Maruthi, M. N., Colvin, J., Seal, S. E. and Muniyappa, V., 2005, Host range, vector relationships and sequence comparison of a begomovirus infecting hibiscus in India. Ann. Appl. Biol., 147: 15-25.

Reddy, M.K., 2010, Climate change and virus diseases of horticultural crops. In Challenges of climate change: India Horticulture, Westville Publishing House, New Delhi, pp. 210.

Rekha, A.R., Maruthi, M. N., Muniyappa, V. and Colvin, J., 2005, Occurrence of three genotypic clusters of Bemisia tabaci and the rapid spread of the B biotype in south India. Entomologia Experimentalis et Applicata, 117: 221–233.

Sarra, S., Oevering, P., Guindo, S. and Peters, D., 2004, Wind-mediated spread of Rice yellow mottle virus (RYMV) in irrigated rice crops. Plant Pathol. 53, 148–153.

Shankarappa, K. S., Rangaswamy, K. T., Raghavendra, N., Aswatha Narayana, D.S., Lakshminarayana Reddy, C.N., Prameela, H. A., Colvin, J., and Muniyappa, V., 2006, Monitoring for B biotype in certain whitefly (Bemisia tabaci) populations collected from different locations in India. Indian Journal of Virology, 17: 127-128.

Southerst, R.W., Constable, F., Finlay, K.J., Harrington, R., Luck, J. and Zalucki, M.P., 2011, Adapting to crop pest and pathogen risks under changing climate. Wiley Interdiscip. Rev. Clim. Change 2, 220–237.

Tomlinson, J. A., 1987, Epidemiology and control of virus diseases of vegetables. Ann. Appl. Biol., 110: 661-81.

Varma, A. and Malathi, V. G., 2003, Emerging Gemini virus problems. A serious threat to crop production. Ann. Appl. Biol., 142: 145-64.

Wang, G.-H., Wang, X.-X., Sun, Y.-C. and Ge, F., 2014, Impacts of elevated CO_2 on *Bemisia tabaci* infesting Bt cotton and its parasitoid *Encarsia formosa*. Entomol. Exp. Appl. 152, 228–237.

Yeates, G.W. and Newton, P.C.D., 2009, Long term changes in topsoil nematode populations in grazed pasture under elevated atmospheric carbon dioxide. Biol. Fertil. Soils, 45, 799–808.

Yeates, G.W., Newton, P.C.D. and Ross, D.J., 2003, Significant changes in soil micro fauna in grazed pasture under elevated carbon dioxide. Biol. Fertil. Soils, 38, 319–326.

Zavala, J.A., Nabity, P.D. and DeLucia, E.H., 2013, An emerging understanding of mechanisms governing insect herbivory under elevated CO_2. Annu. Rev. Entomol. 58, 79–97.

Zhang, S., Li, X., Sun, Z., Shsao, S., Hu, L., Ye, M., Zhou, Y., Xia, X., Yu, J. and Shia, K., 2015, Antagonism between phytohormone signaling underlies the variation in disease susceptibility of tomato plants under elevated CO_2. J. Exp. Bot., 66 (7), 1951–1963.

9

Pest Forecasting

Y.G. Prasad

Introduction

Insects and crops have coevolved over a long period of time. As a result insects have adapted themselves to a particular crop or group of crop plants and also have specialized feeding niches. Pest forewarnings provide lead time for impending attacks and thus minimise crop loss and optimize pest control leading to reduced costs of cultivation. Pest forecasts will have to take several factors into consideration while being devised in the form of model outputs and disseminated as alerts to farmers. Many pests and diseases are regular in their appearance at susceptible stages of crop growth *e.g.*, pod borer, *Helicoverpa armigera* (Hubner) on pigeonpea at flowering period. The pest appears year after year with its population fluctuating around a seasonal mean. Some pests and diseases are endemic to certain locations (*e.g.*, rice gall midge) while some are cyclical or sporadic in their appearance (*e.g.*, Bihar hairy caterpillar, *Spilosoma obliqua* Walker on sunflower). Some are characterized as outbreak pests or diseases because of their potential to rapidly increase in their abundance in a short span triggered by certain congenial factors leading to widespread epidemics *viz.*, brown plant hopper, *Nilaparvatha lugens* Stal and blast in rice; *Spodoptera litura* (Fab.) on soybean and cotton; mealybug, *Paracoccus marginatus* Williams and Granara de Willink on papaya and several other crops. Species which have a wider host range have better chances of carry over to the next season or year. Species which are monophagous or oligophagous have resting or dormant phases in their lifecycles to tide over unfavourable weather condtions or lack of a susceptible host or crop stage. Pest and disease appearance on a crop follows its life cycle and is also intricately connected to the crop phenology. Some insects complete only a single generation a year (*e.g.* red hairy caterpillar, *Amsacta albistraga* Walker) while others have multiple and often overlapping generations.

Pest forecasts need to consider these intrinsic attributes of the insect pest while determining enviromental and host factors.

Factors Influencing Pest Abundance

Pest populations fluctuate in timing and intensity of their occurrence depending on location and season. Mostly they tend to fluctuate over a mean level. This average population over time when computed across several years results from the sum of action of all positive and negative factors influencing pest populations. Pests of host plants in undisturbed habitats such as forestry have their natural cycles in response to their ecosystem interactions and are most likely to attain equilibrium points in their population levels. Pests of agro-ecosystems fluctuate within and between short spans of time *i.e.*, season and off-season, as they experience rapidly changing environments– due to changes in cropping systems and a host of management interventions. As a result crop pests show greater degree of instability in population levels.

An abnormal increase in the numbers of a species on any crop in a short span of time over a large area leads to a significant impact on production and productivity and is considered as a pest outbreak or epidemic. Many factors are responsible for such an outbreak. Foremost among them is the prevalence of congenial weather conditions for insect multiplication and rapid buildup. Growing susceptible cultivars of crop plants in mono cultures on a large scale year after year may help insect pests to multiply and spread fast. Sometimes indiscriminate use of insecticides for pest control disrupts the ecological balance between phytophagous insects and their natural enemies leading to pest outbreaks apart from causing pest resurgence and pesticide resistance. Invasive pest species exhibit outbreak tendency due to lack of natural enemies which are crucial in regulation of pest populations. In general, insects are influenced by extrinsic factors such as weather and crop (Table 1).

Table 1: Weather Factors Influencing Pest Population Dynamics

Factors	*Parameter*	*Influence on*
Weather	Temperature (maximum and minimum)	Insect developmental rate, survival, fecundity, diapause
	Humidity (maximum and minimum)	Survival, egg hatch
	Rainfall (amount and distribution)	Adult emergence, oviposition, spread, mortality (wash-out)
	Bright sunshine hours/Photoperiod	Food quality, oviposition, diapause
	Wind speed and direction	Dispersal and migration
	Microclimatic parameters (canopy temperature, humidity, leaf wetness, soil temperature)	Development and rate of spread, mortality Plant diseases: spore germination, infection and sporulation
Crop	Genotype	Rate of development, survival
	Phenology	Pest appearance and intensity
	Crop diversity, geometry &stand	Insect diversity, abundance, survival and dispersal
	Nutrition	Fecundity
	Time of sowing	Pest onset
	Cropping system practices	Off-season survival, carry over
	Density dependent mortality factors	Pest regulation

On a global scale, seasonal temperatures and rainfall patterns constitute major factors that determine the distributions of organisms in space (Birch, 1957). Tropical insects on the average have the same annual variability as insects from temperate zones, but insect populations from dry areas, temperate or tropical, tend to fluctuate more than those from wet areas (Wolda, 1978).

In nature pests are regulated by their natural enemies: parasitoids, predators and pathogens which are again influenced by bio-physical factors (Thomson *et al.*, 2010, Hence *et al.*, 2007). Therefore, precise understanding of population dynamics can result from comprehensive ecological studies. However, despite best efforts gaps in pest ecological databases exist due to the complexity of interactions among the ecosystem components. It is important to realize that weather is only one of the many driving variables that influence pest dynamics.

Modeling Pest Dynamics

Pest prediction models are mathematical descriptions of biological or empirical data as influenced by the environment. They summarize biophysical relationships in the form of equations and provide a basis to test the significance of this relationship. Important points for consideration in any model development are: the level of detail at which a given model is to be developed as the level of detail is linked to the objective and data availability to develop and run the model. Models can range from strictly empirical to most complex and sophisticated descriptive models. A model may be discrete or continuous, static or dynamic, and deterministic or stochastic.

Building and running pest forecast models require access to weather and climate data, in addition to pest and plant data. Models usually require as inputs, measurements of temperature, rainfall and humidity, although other variables may be required either as direct inputs or in computing values for variables not measured. Weather variables need to be measured at the field level, at regional stations, or on a broader scale depending on the need. For many farm management actions, data representative of the field conditions are expected and hence data is taken from automatic weather stations or the nearest observatory.

There are several approaches adopted by modeling groups to achieve the common aim of providing pest forecasts. These approaches broadly fall into two categories: empirical and process-based. Empirical approaches involve estimating pest and disease incidence and intensity through experimentation and surveys on crops not subjected to control interventions and establishing relationships with concurrent, prevailing weather and/or past weather factors. These studies could be conducted at single stations in which the emphasis is on delineation of differences in meteorological conditions in epidemic and non-epidemic years or multi-station studies in which the emphasis is on delineation of meteorological conditions leading to changes in periods and intensity of infestations. A multi-station study is preferred as it facilitates corroboration of the general surmises and leads to maximization of data in a short period if observations are recorded on crop stands sown at periodic intervals at a number of stations (Venkataraman and Krishnan, 1992). It should be noted that findings from empirical field studies can straight away be applied in climatologically analogous areas but can give misleading results when applied to other areas.

Process-based models are essentially descriptive or mechanistic and are dependent on insect life cycles, ecological and physical processes. These models are complex to build, sophisticated to run and are essentially computer based. Correlation and regression approach to modeling has been widely adopted in the tropics in Asia. Phenology and processed based modeling is in vogue in temperate and tropical regions of developing countries.

Regression Models

Regression models consider pest/disease incidence variables as dependent and suitable independent variables such as weather variables, crop stage; population of natural enemies/predators *etc.* is used. The form of the model is:

$$Y = \beta 0 + \beta 1\ X1 + \beta 2X2 + \dots\dots\dots + \beta pXp + e$$

where β0, β1, β2,.......βp are regression coefficients, X1, X2, ...,Xpare independent variables and e is error term. These variables are used in original scale or on a suitable transformed scale such as cos, log, exponential *etc.*

In case of regression models based on weather indices, using weekly and fortnightly weather variables, suitable indices are worked out which are used as regressors in the model. If information on favourable weather conditions is known, subjective weights based on this information can be used for constructing weather indices. In the absence of such information correlation coefficients between Y and respective weather variable/product of weather initial & final periods for which weather data were included in the model and e is error term (Chattopadhyay *et al.*, 2005a, Chattopadhyay *et al.*, 2005b, Desai *et al.*, 2004 and Dhar *et al.*, 2008).

In regression analysis, the unfitted errors between a regression model and observed data are generally assumed as observation error that is a random variable having a normal distribution, constant variance, and a zero mean. However, in fuzzy regression analysis, the same unfitted errors are viewed as the fuzziness. Fuzzy regression can be quite useful in estimating the relationship among variables where the available data are imprecise and fuzzy. Fuzzy regression method is based on minimizing fuzziness as an optimal criterion, which can be achieved by linear programming procedures.

Forewarning models can be developed using the principal component technique as normally relevant weather variables are large in number and are expected to be highly correlated with each other. Using the first few principal components of weather variables as independent variables forecast models can be developed. Discriminant function analysis is applied to time series data on weather variables. For this analysis, a series of data for 25-30 years are required. Based on the pest and diseases variables, data can be divided into different groups – low, medium and high *etc.* and using weather data in these groups, linear or quadratic discriminant functions can be fitted which can be used to find discriminant scores. Considering these discriminant scores as independent variables and diseases/pest as dependent variable, regression analysis can be performed. Johnson *et al.* (1996) used-discriminant analysis for forecasting potato late blight.

Machine Learning Techniques

Machine learning techniques offer many methodologies like decision tree induction algorithms, genetic algorithms, neural networks, rough sets, fuzzy sets as well as many hybridized strategies for the classification and prediction. Decision tree induction represents a simple and powerful method of classification that generates a tree and a set of rules, representing the model of different classes, from a given dataset. Decision Tree (DT) is a flow chart like tree structure, where each internal node denotes a test on an attribute, each branch represents an outcome of the test and each leaf node represents the class. The top most nodes in a tree are the root node. One of the strengths of decision trees compared to other methods of induction is the ease with which they can be used for numeric as well as non-numeric domains. Another advantage of decision tree is that it can be easily mapped to rules. A decision tree analysis for predicting the occurrence of the pest, *Helicoverpa armigera* and its natural enemies on cotton based on economic threshold level was developed (Pratheepa *et al.*, 2011). The data mining process for construction of decision tree and set of rules is shown in Figure 1. The occurrence of cotton bollworm *H. armigera* (Hübner) was greatly influenced by its natural enemies, *viz.* spiders and *C. carnea* and by abiotic factors. In this analysis, population dynamics of the pest and its natural enemies was studied using Shannon information measure with decision tree induction approach. The developed classification model has the ability to successfully treat 'categorical' variables as well as 'continuous' variables in the database. Pest incidence was classified into two classes, *viz.*, and low and high based upon economic threshold level (ETL). R2 for low and high class values was 0.66 and 0.21; respectively. Our studies showed that season influenced the population dynamics of H. armigera among all the factors. It was found that the misclassified testing data were 8.8 %. The confusion matrix for the testing set revealed that the classification was done more accurately using the training set. Hence, this approach could be successfully-utilized to understand the role of natural enemies and weather factors on the occurrence of *H. armigera* (Hübner) as well as prediction of this pest.

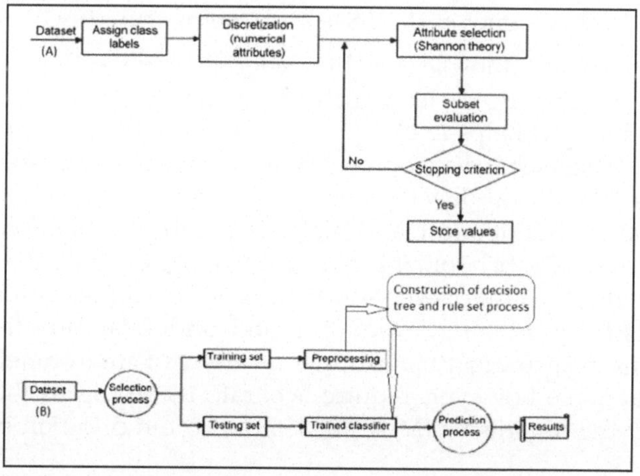

Figure 1: (A) Data mining process for construction of decision tree; (B) Training and testing process (Pratheepa *et al.*, 2011).

Artificial Neural Networks (ANNs) is another attractive tool under machine learning techniques for forecasting and classification purposes. ANNs are data driven self-adaptive methods in that there are few apriori assumptions about the models for problems under study. These learn from examples and capture subtle functional relationships among the data even if the underlying relationships are unknown or hard to describe. After learning from the available data, ANNs can often correctly infer the unseen part of a population even if data contains noisy information. As forecasting is performed via prediction of future behaviour (unseen part) from examples of past behaviour, it is an ideal application area for ANNs, atleast in principle (Kumar *et al.*, 2010). However, the technique requires a large data base.

Insect Development and Phenology Models

Insects are incapable of their own internal temperature regulation and hence their development depends on the temperature to which they are exposed. The studies of population dynamics often involve modeling growth as a function of temperature. The rate summation methodology has perhaps proven to be the most viable approach to such modeling (Stinner *et al.*, 1974). The most common development rate model, often called degree-day summation, assumes a linear relationship between development rate and temperature between lower and upper development thresholds (Allen, 1976). This method works well for optimum temperatures (Ikemoto, 2005). The linear model assumes that rates are proportional to temperature, and since amounts are integrals of rates, the amount of development is the integral of temperature (or a linear function of it) along a time axis and has units of temperature-time (*e.g.*, degree-days). Temperature-dependent development in insects can be approached using either developmental time or rate of development. Rate of development is traditionally utilized because rate models were created from biochemical and biophysical properties (Sharpe and DeMichele, 1977). Most of the earlier models failed to take into consideration variation between individual insects in their rate of development that is responsible for spread of activity of a pest (Regniere, 1984, Wagner *et al.*, 1984a and Phelps *et al.*, 1993).

Insect species that exhibit seasonality generally have resting phases, diapause or aestivation, in their life cycles which can be accommodated in Monte Carlo simulation modeling (Phelps *et al.*, 1993). Since some temperatures are lethal to organisms, it is obvious that development must be a nonlinear temperature function at the extremes. Non-linear development rate functions based on enzyme kinetics were developed to describe low and high temperature inhibition (Stinner *et al.*, 1974, Logan *et al.*, 1976, Schoolfield *et al.*, 1981, Wagner *et al.*, 1984b). Phenology models help predict the time of events in an insects' development and are important analytical tools for predicting, evaluating and understanding the dynamics of pest populations in agro-ecosystems under a variety of environmental conditions. Accurate predictions, however, require accurate recording of the temperatures experienced by the organisms (Morgan, 1991) also the duration of development (Danks, 2000).

Degree-day models (Higley *et al.*, 1986) have long been used as part of decision support systems (DSS) to help growers predict spray timing or begin pest scouting

(Welch *et al.*, 1978). Phenology models are also used as one component of risk analysis for predicting exotic pest establishment (Baker, 1991; Jarvis and Baker, 2001). A well-known example is the DYMEX modeling package (Yonow *et al.*, 2004). Some other modeling packages, for example CLIMEX although not strictly phenology models may also use some developmental requirements for risk assessment (Sutherst *et al.*, 1991, 1999, 2000). Another example is the web-based North Carolina State University-APHIS Plant Pest Forecast (NAPPFAST) modeling system that links daily climate and historical weather data with biological models to produce customized risk maps for phytosanitary risk assessments (Borchert and Magarey, 2005). Resources like the Crop Protection Compendium (CAB International, 2004) often summarize insect development, while the University of California Statewide IPM program lists development data for insects on their website (http://www.ipm.ucdavis.edu/MODELS) for use in DD models. An Insect Development Database containing the developmental requirements for over 500 insect species was created (Nietschke *et al.*, 2007). Insect Life Cycle Modeling software (ILCYM), a generic open source computer aided tool, facilitates the development of phenology models and prediction of pest activity in specific agroecologies (Sporleder *et al.*, 2009).

Pest Simulation Models and Decision Support Systems

Simulation models based on mathematical description of biological data as influenced by environment are more easily applied across locations and environments. Computer programs or software to run these models facilitates the practical application of these models in understanding population dynamics and dissemination of pest forecasts for timely pest management decisions (Coulson and Saunders, 1987) (Table 2). Simulation approach offers flexibility for testing, refinement, and sensitivity analysis and field validation of developed models over a wide range of environmental conditions. Thorough descriptions of cropping systems being managed or studied are needed to explain the interactions among pests, plants, and environment (Colbach, 2010). Systems models or other prediction schemes can be used with appropriate biological, environmental, economic, or other inputs to analyze the most effective management actions, based on acceptable control, sustainability, and assessment of economic or other risks (Strand, 2000).

Crop Growth Simulation Models Coupled with Pest Sub-Routines

Crop system models can be used to generate information on the status of the crop, its pests, and its environment under different scenarios, including different management options. In practice, there are few examples of these models that include all the necessary components and can be used for practical decision making. However, a more practical approach has been the development of individual crop and pest components that can be analyzed at the same time to give information that can improve decisions.

Development of decision support system for agro-technology transfer (DSSAT 4 funded by USAID) allowed rapid assessment of several agricultural production systems around the world to facilitate decision-making at the farm and policy levels. The trend in development of crop system models is to go for modular approach (www.icasa.net). Development of stand-alone decision support systems for pest components could lead to their practical use. In developed countries, dynamic websites that include interactive models, GIS (Geographical Information System) based decision systems, real-time weather, and market information are being rapidly developed and made available on the Internet (www.effita.net) enabling farmers for real time use in crop management.

Table 2: Prediction Models Developed under the National Agricultural Technology Project (NATP) and National Agricultural Innovation Project of ICAR

Model/Tool	*Crop*	*Pest*	*Reference*
Life cycle modeling	Cotton	Mealybug	Prasad *et al.*, 2012
Life cycle modeling	Rice	Leaf folder	Padmavathi *et al.*, 2013
CART and NN models	Rice	Yellow stem borer	Kumar *et al.*, 2012
Regression	Cotton	Pink bollworm	Vennila *et al.*, 2011
Artificial neural network	Mustard	Powdery mildew	Ratna Raj Laxmi and Amrender Kumar, 2012
Regression	Pigeonpea	Pod borer	Vishwa Dhar *et al.*, 2008
Degree-day model	Mustard	Aphid	Chakravarthy and Gautam, 2004
Regression	Mustard	Aphid	Chattopadhyay *et al.*, 2005
Regression	Rice	Rice blast	Yella Reddy *et al.*, 2006
Epidemiological, Regression	Mustard	Alternaria blight	Chattopadhyay *et al.*, 2005
Online decision tools	Generic	Generic	http://www.crida.in/naip/comp4/dss_pest.html
Online forecast tool	Mustard	Aphid	Vinod Kumar *et al.*, 2012 (www.drmr.res.in/aphidforecast/models.html)
Online forecast tool	Rice	Blast	Kaundal *et al.*, 2006 (www.imtech.res.in/raghava/rbpred/)

Conventional approaches of using empirical models for quantifying yield losses due to crop pests are limited in their scope and application since these equations are location and data-specific and insensitive to variable cropping and pest conditions. Crop growth models provide a physiologically based approach to simulate pest damage and crop interactions. There have been many efforts to use crop growth models to simulate the effect of pest damage on crop growth and yield by linking the damage effect of pest population levels to physiological rate and state variables of these models. The insect pest-crop modelling has been discussed in detail by Boote *et al.* (1983); Coulson and Saunders (1987). A distribution delay model including attrition was applied to simulate population changes of rice leaf-folders. Bases on metabolic pool approach, leaf folder feeding and hence leaf mass losses to the rice plant were described with a generalized functional response model which is 'source' and 'sink' driven (Graf *et al.*, 1992). Furthermore, this model stresses the influence of adult migration and natural enemies on leaf-folder population dynamics, both significant and poorly investigated aspects of the leaf-folder life cycle. Later, a generic approach to simulate the damage effects of single or multiple pests attempted using crop growth models such as CERES-rice (which is a part of DSSAT) in Philippines (Pinnschmidt *et al.*, 1995) and Info Crop in India (Reji *et al.*, 2008). Pest damage levels from field scouting reports can be entered and damage is applied to appropriate physiological coupling points within the crop growth model including leaf area index, stand density, intercepted light, photosynthesis, assimilate amount and translocation rate, growth of different plant organs and leaf senescence. Equations and algorithms were developed to describe competition among multiple pests and to link the computed total damage to the corresponding variables in the crop models. These approaches provide a basis to explore dynamic pest and crop interactions in determining pest management strategies which minimize yield losses.

Operational Pest Forecasting

Farmers are mainly interested in current disease and pest severity data, preferably for their localities to aid their decision-making in crop protection. Pest monitoring data along with complementary monitoring of weather data is crucial to run pest forecast models and make available forecasts for operational use. Weather measurements under field conditions from several geo-referenced sites in the crop cultivated regions additionally provides spatial information which can be used for generating pest forecast maps (Huang *et al.*, 2008). In Bayern (Germany) a measuring network of 116 field weather stations is used to estimate the development of pests in relation to weather requirements based on forecast models and computer-based DSS for near real time dissemination to farmers (Tischner, 2000). The results of crop and horticulture specific models and DSS are supplemented by field-monitoring data which then serve as the main input for the warning services and cost effectively disseminated through the Internet (Bugiani *et al.*, 1996; Jorg, 2000). A computerized national forecasting network in apple orchards transmits data from the field to system headquarters automatically. The national forecasting network in Turkey has been expanded and covered apple orchards in 34 provinces in 2006, using 115 electronic forecasting and warning stations (Atlamaz *et al.*, 2007).

In India, a grommet network under the All India Coordinated Research Project (AICRPAM) provides agro-advisories on http://www.cropweatheroutlook.org. An online tool has been developed for forecasting of mustard aphid (*Lipaphis erysimi*) from incidence data on mustard cultivars sown at staggered intervals at multiple locations over a period of 3 years. Multiple step-wise regression models were built between aphid incidence over time and composite weather variables with correlation coefficients as weights. Weather indices were computed for different weeks after sowing until the forecast was provided. Using SAS statistical software, models were devised for forecasting 1) crop age at first appearance of aphid on crop 2) crop age at highest aphid population (aphid counts on 10 cm of terminal shoot) in the season and 3) Peak aphid population on the crop in the season for different locations (Vinod Kumar *et al.*, 2012).

'RB-Pred' a web-based server, was developed for forecasting leaf blast severity in rice based on the weather variables (Kaundal *et al.*, 2006). 'RB-Pred' predicts rice leaf blast severity (%) based on the weather parameters input by the user. It uses the regression module called support vector regression (SVR), a powerful machine learning technique called support vector machine (SVM). The SVM learns how to classify from a training set of feature vectors, whose expected outputs are already known. The training enables a binary classifying SVM to define a plane in the feature space, which optimally separates the training vectors of two classes. When a new feature vector is inputted, its class is predicted on the basis of which side of the plane it maps.

Conclusion

Pest monitoring is the foundation for issue of early warnings, development and validation of pest forecast models and decision support systems which are crucial for design and implementation of successful IPM programmes. Models are useful ways of synthesizing the available information and knowledge on population dynamics of pests in agro-ecosystems and natural habitats. However, any model needs to be tested and validated prior to use. Development of long-term monitoring on a spatial scale on crop-pest-weather relationships will narrow the gaps in knowledge required for reliable forecasts. In the last decade, two ICAR projects were funded by the World Bank and implemented with CRIDA as the consortium lead in the area of pest forecast research in key field crops such as rice, cotton, groundnut, sugarcane, pigeonpea and mustard. Significant contributions were made under the NATP project and useful leads are forthcoming under the ongoing NAIP project (Table 2). Computer-based systems have increased the speed and accuracy of forecasting as well as decreasing its costs. Recent developments in information and communication technology offer great scope for wide dissemination and use of pest forecasts. In the tropics, agro-ecosystems are characterized by greater crop diversity in small parcels of land and dynamically changing weather. Available generic simulation models need to be tested with location specific inputs for greater accuracy. In developing countries there is a strong need to establish functional networks for specific crop sectors with the major objective of pest forecasting through models and decision support systems.

References

Allen I C 1976. A modified sine wave method for calculating degree days. Environmental Entomology., 5 (3), 388-396.

Atlamaz A, Zeki C and Uludag 2007. A. The importance of forecasting and warning systems in implementation of integrated pest management in apple orchards in Turkey. EPPO Bulletin., 37(2), 295-299.

Baker C R B 1991. The validation and use of a life-cycle simulation model for risk assessment of insect pests. EPPO Bull., 21, 615–622.

Birch L C 1957. The role of weather in determining the distribution and abundance of animals. Cold Spring Harbor Symp. Quant. Biol., 22, 203-18.

Boote K J, Jones J W, Mishoe J W and Berger R. D 1983. Coupling pests to crop growth simulators to predict yield reductions. *Phytopathology.*, 73, 1581-1587.

Borchert D M and Magarey R D 2005. A guide to the use of NAPPFAST./http://www.nappfast.org/usermanual/nappfast-manual.pdfS.

Bugiani R, Tiso R, Butturini A, Govoni P and Ponti I. 1996. Forecasting models and warning services in Emilia-Romagna (Italy). EPPO Bulletin., 26(3-4), 595-603.

CAB International Crop Protection Compendium 2004. CAB International, Wallingford, UK.

Cannon R J C, Koerper D and Ashby S 2004. Gypsy moth, *Lymantria dispar*, outbreak in northeast London, 1995-2003. International Journal of Pest Management., 50(4), 259-273.

Chakravarty N V K and Gautam R D 2004. A degree-day based forewarning system for mustard aphid, *Lipaphis erysimi* (Kaltenbach). Journal of Agrometeorology, 5 : 215-222.

Chander S, Kalra N and Aggarwal P K 2007. Development and application of crop growth simulation modelling in pest management. Outlook on Agriculture., 36(1), 63-70.

Chattopadhyay C Agrawal R Kumar A Bhar L M Meena P D Meena R L Khan S A Chattopadhyay A K Awasthi R P Singh S N Chakravarthy N V K Kumar A Singh R B and Bhunia C K (2005a). Epidemiology and forecasting of Alternaria blight of oilseed Brassica in India – a case study. Zeitschrift fürPflanzenkrankheiten und Pflanzenschutz (Journal of Plant Diseases and Protection), 112 : 351-365.

Chattopadhyay C Agrawal R Kumar Amrender Singh Y P Roy, S K Khan S A Bhar L M Chakravarthy N V K Srivastava A Patel B S Srivastava, B Singh C P and Mehta S C 2005b. Forecasting of *Lipaphis erysimi* on oilseed Brassicas in India – a case study, Crop Protection, 24 : 1042-1053.

Colbach N 2010. Modelling cropping system effects on crop pest dynamics: how to compromise between process analysis and decision aid. Plant Science., 179(1-2), 1-13.

Coulson R N and Saunders M C 1987. Computer-assisted decision making as applied to entomology. Annual Review of Entomology., 32, 415-437.

Danks H V 2000. Measuring and reporting life-cycle duration in insects and arachnids. European Journal of Entomology., 97, 285-303.

Dillon M L and Fitt G P 1990. HEAPS; a regional model of Heliothis population dynamics, pages 337-344 In: Proceedings of fifth Australian Cotton Conference 8-9 August, 1990, Broadbeach Queensland, Brisbane, Queensland, Australia: Australian Cotton Grower's Research Association.

Drake V A, Wang H K and Harman I T 2002. Insect monitoring radar: remote and network operation. Computers and Electronics in Agriculture., 35, 77-94.

Feng H, Wu X, Wu B and Wu K 2009. Seasonal migration of *Helicoverpa armigera* (Lepidoptera: Noctuidae) over the bohai sea. Journal of Economic Entomology. 102(1), 95-104.

Fitt G P, Dillon M L and Hamilton J G 1995. Spatial dynamics of *Helicoverpa* populations in Australia: Simulation modeling and empirical studies of adult movement. Agricultural ecosystem and environment., 13, 177-192.

Higley L G, Pedigo L P and Ostlie K R 1986. DEGDAY: a program for calculating degree-days, and assumptions behind the degree-day approach. Environ. Entomol., 15, 999–1016.

Hopkinson R F and Soroka J J.Air trajectory model applied to an in-depth diagnosis of potential diamondback moth infestations on the Canadian prairies. Agricultural and Forest Meteorology.

Huang Y, Lan Y, Westbrook J K and Hoffmann W C 2008. Remote Sensing and GIS Applications for Precision Area-Wide Pest Management: Implications for Homeland Security. The GeoJournal Library, 94: 241-255.

Ikemoto T 2005. Intrinsic optimum temperature for development of insects and mites. Environmental Entomology, 34 : 1377–1387.

Jarvis C H and Baker R H 2001. A risk assessment for non-indigenous pests, I. Mapping the outputs of phenology models to assess the likelihood of establishment. Diversity Distrib. 7, 223–235.

Jorg E 2000. Structure and organization of warning services in Rheinland-Pfalz (Germany). EPPO Bulletin., 30(1), 31-35.

Jorg E, Racca P, Preib U, Butturini A, Schmiedl J and Wojtowicz A 2007. Control of Colorado potato beetle with the simlep decision support system. EPPO Bulletin., 37(2), 353-358.

Kaundal R Kapoor A S and Raghava G P S 2006. Machine learning techniques in disease forecasting: a case study on rice blast prediction, BMC Bioinformatics, 7 : 485.

Klueken A M, Hau B, Ulber B and Poehling H M 2009. Forecasting migration of cereal aphids (Hemiptera: Aphididae) in autumn and spring. Journal of Applied Entomology., 133(5), 328-344.

Knutson, Allen E, Muegge and Mark A 2010. A degree-day model initiated by pheromone trap captures for managing pecan nut casebearer (Lepidoptera: Pyralidae) in pecans. Journal of Economic Entomology., 103(3), 735-743.

Kumar A, Prasad Y G Vennila S Vasantabhanu K Prabhakar M Padmakumari A P K and Katti G 2012. A comparative analysis of classification and regression tree (CART) and neural network (NN) models in prediction of rice yellow stem borer. 14th Annual Conference of Society of Statistics, Computer and Applications organised at the Department of Statistics, Saurashtra University, Rajkot during 24-26 February 2012.

Lewis T 1980. Britain's pest monitoring network for aphids and moths. EPPO Bulletin., 10 (2), 39-46.

Liebhold A M and Tobin P C 2008. Population ecology of insect invasions and their management. Annu. Rev. Entomol., 53, 387–408.

Logan J A, Wollkind D T, Hoyt J C and Tanigoshi L K 1976. An analytic model for description of temperature dependent rate phenomena in arthropods. Environmental Entomology., 5, 1 130-1 140.

Morgan D 1991. The sensitivity of simulation models to temperature changes. EPPO Bulletin., 21(3), 393-397.

Morris R F, Miller C A 1954. The development of life tables for the spruce budworm. Can. J. Zool., 32, 283-301.

Morris R.F Ed 1963. The dynamics of epidemic spruce budworm populations. Can. EntomologistMem., 31, 332.

Nietschke B S, Magarey R D, Borchert D M, Calvin D D and Jones E 2007. A developmental database to support insect phenology models. Crop Protection, 26, 1444-1448.

Padmavathi Ch, Katti, G, Sailaja, V, Padmakumari, A P Jhansilakhsmi, V Prabhakar M and Prasad Y G 2013. Temperature thresholds and thermal requirements for the development of rice leaf folder, *Cnaphalocrocis medinalis*. Journal of Insect Science (In press).

Pinnschmidt H O. Batchelor W D, and Teng P S 1995. Simulation of multiple species pest damage in rice using CERES-Rice. Agricultural Systems 48: 193–222.

Phelps K, Collier R H, Reader R J and Finch S 1993. Monte Carlo simulation method for forecasting the timing of pest insect attacks. Crop Protection., 12(5), 335- 341.

Phelps K, Reader R J and Hinde C J 1999. HIPPO - flexible software for the construction, integration and distribution of biologically realistic models. *Aspects of Applied Biology*, **55**, 81-88.

Prasad N V V S D, Mahalakshmi M S and Rao N H P 2008. Monitoring of cotton bollworms through pheromone traps and impact of abiotic factors on trap catch. Journal of Entomological Research., 32(3).

PrasadY G Prabhakar M Sreedevi G Ramachandra Rao G and Venkateswarlu B 2012. Effect of temperature on development, survival and reproduction of the mealybug, *Phenacoccus solenopsis* Tinsley (Hemiptera: Pseudococcidae) on cotton, Crop protection 39 : 81-88.

Pratheepa M Meena K Subramaniam K R Venugopalan R and Bheemanna H 2011. A decision tree analysis for predicting the occurrence of the pest, *Helicoverpa armigera* and its natural enemies on cotton based on economic threshold level. Current Science, 100 : 242-246.

Ratna Raj Laxmi and Amrender Kumar 2011. Forecasting of powdery mildew in mustard (*Brassica juncea*) crop using artificial neural networks approach, Indian Journal of Agricultural Sciences, 81: 855–60.

Regniere J 1984. A method of describing and using variability in development rates for the simulation of insect phenology. Canadian Entomologist., 116, 1367-1376.

Reji G and Chander S 2008. A degree-day simulation model for the population dynamics of the rice bug, *Leptocorisa acuta* (thunb.). Journal of Applied Entomology., 132(8), 646-653.

Reji G, Chander S and Aggawal P K 2008. Simulating rice stem borer, *Scirpophaga incertulas* Wlk., damage for developing decision support tools. Crop Protection.

Schoolfield R M, Sharpe P J and Magnuson C E 1981. Nonlinear regression of biological temperature-dependent rate models based on absolute reaction-rate theory. Journal of Theoretical Biology., 88, 719-731.

Sporleder M, Simon R., Gonzales J, Chavez D, Juarez H, De Mendiburu F and Krosche J 2009. ILCYM-Insect Life Cycle Modelling. A software package for developing temperature based insect phenology models with applications for regional and global risk assessments and mapping. International Potato Centre, Lima, Peru. pp. 62.

Stinner R E, Gutierrez A P and Butler G D Jr 1974. An algorithm for temperature-dependent growth rate simulation. Canadian Entomologist., 106, 519-524.

Strand J F. Some agro-meteorological aspects of pest and disease management for the 21st century.

Sutherst R W, Maywald G F, Yonow T and Stevens P M 1999. CLIMEX: Predicting the Effects of Climate on Plants and Animals. CSIRO Publishing, Collingwood, Australia, 88pp.

Sutherst R W and Maywald G F 1985. A computerised system for matching climates in ecology, *Agriculture Ecosystems and Environment*, **13** : 281-99.

Sutherst R W and Maywald G F 2005. A climate-model of the red imported fire ant, *Solenopsis invicta* Buren (Hymenoptera: Formicidae): implications for invasion of new regions, particularly Oceania, *Environmental Entomology* 34 : 317-335.

Sutherst RW, Maywald G F and Russell B L 2000. Estimating vulnerability under global change: modular modelling of pests. Agric. Ecosyst. Environ., 82, 303–319.

Sutherst RW, Maywald G F and Bottomley W 1991. From CLIMEX to PESKY, a generic expert system for pest risk assessment. Bull. OEPP., 21 : 595–608.

Thomson L J, Macfadyen S and Hoffman A A 2010. Predicting the effect of climate change on natural enemies of agricultural pests. Biological Control, 52 : 296-306.

Tischner H 2000. Instruments of the warning service for plant protection in Bayern (Germany). EPPO Bulletin, 30 : 103-104.

Trnka M, Muska F, Semeradvoa D, Dubrovsky M, Kocmankova E and Zalud Z 2007. European corn borer life stage model: Regional estimates of pest development and spatial distribution under present and future climate. Ecological modeling, 207: 61-84.

Varley G C and Gradwell G R 1970. Recent advances in insect population dynamics. Annual Review of Entomology, 15: 1-24.

Venkataraman S and Krishnan A 1992. Weather in the incidence and control of pests and diseases. In: Crops and Weather (Eds Venkataraman S and Krishnan A) Publications and Information Division, Indian Council of Agricultural Research, New Delhi. pp. 259-302.

Vennila S Meenu Agarwal Dharmendra Singh Prasenjit Pal and Biradar V K 2011. Approaches to weather based prediction of insects: a case study on cotton pink bollworm, *Pectinophora gossypiella*, Indian Journal of Plant Protection, 39 : 163-169.

Vinod Kumar Amrender Kumar and Chattopadhyay C 2012. Design and implementation of web-based aphid (*Lipaphis erysimi*) forecast system for oilseed Brassicas, Indian Journal of Agricultural Sciences, 82: 608–14.

Dhar V, Singh S K Trivedi T P Das D K Choudhary R G and Kumar M 2008. Forecasting of *Helicoverpa armigera* infestation on long duration pigeonpea in central Uttar Pradesh. Journal of Food Legumes 21: 189–92.

Wagner T L, Wu H I, Sharpe P J H and Coulson R N 1984a. Modeling distributions of insect development time: a literature review and application of the Weibull function. Annals of the Entomological Society of America., 77 : 475-487.

Wagner T L, Wu H I, Sharpe P J H, Schoolfield R M and Coulson R N 1984b. Modclling insect development rates; a literature review application of a biophysical model, Annals of the Entomological Society of America., 77: 208-225.

Welch S M, Croft B A and Michels M F 1978. PETE: an extension phenology modeling for management of muti-species pest complex. Environmental Entomology, 7: 487-494.

Willocquet L, Savary S, Fernandez L, Elazegui F A, Castilla N, Zhu D, Tang Q, Huang S, Lin X, Singh H. M and Srivastava R K 2002. Structure and validation of RICEPEST, a production situation-driven, crop growth model simulating rice yield response to multiple pest injuries for tropical Asia. Ecological Modelling., 153 : 247–268.

Witzgall P, Kirsch P, Cork A 2010. Sex pheromones and their impact on pest management. Journal of chemical ecology., 36 (1), 80-100.

Wolda H 1978. Fluctuations in abundance of tropical insects. American Naturalist, 112: 1017-45.

Yella Reddy D Prabhakar M Ramakrishna Y S Reddy C S Prasad Y G and Nagalaksmi T 2006. Dynamic cumulative weather based index for forewarning of rice blast, Journal of Agro-meteorology, 8 : 1-6.

Yonow T, Zalucki M P, Sutherst R W, Dominiak B C, Maywald G F, Maelzer D A and Kriticos D J 2004. Modelling the population dynamics of the queensl and fruit fly, Bactrocera (Dacus)tryoni: a cohort-based approach incorporating the effects of weather. Ecological modeling. 17, 39-30.

Zalucki M P and Furlong M J 2005. Forecasting *Helicoverpa* populations in Australia: A comparison of regression based models and a bioclimatic based modeling approach. Insect Science., 12 (1), 45-46.

Zou L, Stout M J and Ring D R 2004. Degree-day models for emergence and development of the rice water weevil (Coleoptera: Curculionidae) in Southwestern Louisiana. Environmental Entomology, 33: 1541-1548.

10

Forecasting and Development of Prediction Models for Plant Disease Management

Y.S. Amaresh, D.S Aswathanarayana and Gururaj Sunkad

Introduction

Epidemiology is primarily concerned with epidemics (epiphytotics), but the term has a wide meaning and has come to include most field aspects of plant diseases it is the interaction of pathogen, plant and environment, populations of plants and pathogens rather than individuals. Epidemiology covers the effects of environmental factors on disease prevalence, incidence and severity, in addition to survival and spread of plant pathogens.

The word epidemic comes form the greek word mean epi = on and demons = people literally epidemic as defined as affecting or tending to affect many individuals with in a population, a community or a region at the same time. It is outbreaks a sudden and rapid growth spread and development. Epidemiology can defined as the study of disease in populations or it as the study of environmental factors that influence the amount and distribution of disease in population or it as the study of rates of change in the amount of disease in time, in space or both (Van der Plank 1963; Cowling and Horsfall, 1978).

The disease is on output of the interaction among host, pathogen and the environment over a period of time given that a susceptible host and an aggressive pathogen are available, weather determines that intensity of the disease over time. To capture the dynamics of the interactions among there biophysical variables efforts have been made in past to understand the variables and the interaction through modelling approaches

Disease Forecasting

Forecasting of plant diseases means predicting for the occurrence of plant disease in a specified area ahead of time, so that suitable control measures can be undertaken in advance to avoid losses. Disease forecasts are predictions of probable outbreaks or increase in intensity of disease. It involves well organized team work and expenditure of time, energy and money. It is used as an aid to the timely application of chemicals. Among the first spray warning services to be established for growers, were the grapevine downy mildew forecasting schemes in France, Germany and Italy in the 1920s. Disease forecasting methods are available for the following plant diseases.

Sl.No.	*Plant disease*	*Countries*
1	Grapevine downy mildew	Australia, France, Germany, Greece, Italy, Romania, Spain, USSR, Yugoslavia
2	Cucurbit downy mildew	U.S.A.
3	Potato late blight	Australia, Brazil, Finland, France, Germany, Greece, Japan, Netherlands, Norway, Peru, U.K, and U.S.S.R.
4	Tobacco blue mould	Canada, U.S.A.
5	Apple and pear scab	Australia, Canada, Netherlands, New Zealand, U.S.A
6	Sugarbeet root rot (*Aphanomyces* sp.)	U.S.A
7	Wheat brown (Leaf) rust	U.S.A.
8	Corn bacterial wilt (*Erwinia stewartii*)	U.S.A
9	Sugarbeet curly top	U.S.A.

Shrum, 1978, has listed three major research for our interest in forecasting

1. Forecasting provider a means for determining of when and where a given management practices should be applied.
2. Forecasting helps us to identity the gaps that remain in our knowledge, Forecasting provider a systematic way to asses our knowledge of environmental factors and their influence on the progress of disease.
3. Forecasting is also important because it as intellectually stimulating

Requirements or Conditions for Disease Forecasting

There are five main requirements which must be satisfied before a useful and successful disease forecast is made.

1. The disease must cause economically significant damage in terms of yield loss or quality. Damage assessment is essential to develop strategy for controlling a disease. *e.g.*, Annual estimation of yield loss caused by barley powdery mildew (*Erysiphe polygoni*) in England and Wales had ranged from 6 to 13 per cent. Potato late blight can cause a yield loss of 28 per cent if the disease reaches the 75 per cent stage by mid- August.

Diseases like apple scab and potato common scab reduces the quality of the produce lower the value of the harvested crop and cause considerable financial loss to the growers.

2. Control measures must be available at an economically acceptable cost.
3. The disease must vary each season in the timing of the first infections and its subsequent rate of progress. If it does not, there is no need for forecasting.
4. The criteria or model used in making a prediction must be based on sound investigational work carried out in the laboratory and in the field and tested over a number of years to establish its accuracy and applicability in all the locations where its use is envisaged.
5. Growers must have sufficient man power and equipments to apply control measures when disease warning is given. Long-term warnings or predictions are more useful than short- term warning or predictions.

Pest forecasting models or systems are essential tools and pre-requisite of IPM as the insect population and disease incidence levels are deciding factors in pest management. The inter relationship and dependence of plants, pests, natural enemies of pest and weather and their quantification need full understanding. Therefore, forecasting system are developed pest models based on biotic and abiotic factors. These models may be deterministic, statistical, system simulation, expert system, schematic or geographical. These models make the farmers to prepare against pests and decide the timing of application of bio-agents, pesticides and other approaches in integrated pest management. Application of chemical spray against Colorado Beetle and Potato late blight in Soviet Russia based on forecasting model reduced crop losses of about 1.5 million Rubles per year. In Canada the IPM is based on the forecasting information to regulate or economize the intensity and amount of spraying.

The day-degree concept is widely used in USA and. European countries for predicting the levels of attack of pests, relating degree-day with calendar dates for egg hatching, forecasting the optimum moment to destroy adult moths. One of the earliest forecasting work on plant pest was forecasting of late blight of potato (*Phytophthora infestans*) by Dutch rules (Van Everdingen, 1926) which considered several climatic factors together such as dew during night (atleast 4 hours), minimum temperature at night (less than 10°C), mean cloudiness on following day (more than 0.8) and rainfall in the next 24 hours (minimum of 0.1 mm).

In India, one of the earliest works on pest forecasting was started by Dr S Pradhan when he gave the concept of biotic theory of locust population explosion. According to him, temperature was one of the most important factors which determines the duration of various stages in insect life cycle and number of generations. The development of biograph had seen one of the important findings to show the rate of development on different temperature regimes (Pradhan, 1967, 1982). Later, various other models were developed such as wheat stem rust rules (Nagarajan and Singh, 1975,) based on climatologically parameters.

The main requirement for developing pest forecasting models is data on weather parameters, pest population, natural enemies and crop phenology. The data can be generated through regular monitoring and ecological considerations are foundation of pest management programme. Weather parameters have been modelled by multiple regression techniques to assess the per cent incidence of 'Karnal bunt' disease of wheat by Srinivasan and Mavi (1983). The equation developed was:

$$Y = 18.475 - 0.544X1 + 0.108X2 - 0.059X3 - 0.118X4 + 1.192X5 \quad R2 = 0.9066$$

Where,

Y = Karnal bunt disease incidence (%)

X1 = Maximum temperature during 11th week (°C)

X2 = Minimum temperature during 11th week (°C)

X3 = Relative humidity during 11th week (%)

X4 = Cumulative rainfall during 11th week (mm)

X5 = Number of rainy days during 11th week

Requirements for Development of Forewarning System

For development of efficient plant diseases forewarning systems, information about host, pathogen, and environment over a period of time is essential.Each of these components has a significant role to play in development of disease.

1. ***The Pathogen:*** The presence of a pathogen is the main factor that determine occurrence of disease. For each given agro-climatic region, through co-evolution a set of pathogens get adapted and maintain equilibrium. However, due the commercial agriculture in recent years, the balance between host and pathogen are disrupted either through introduction of new crop species or new varieties and hence, the disease scenario changes drastically. In such cases, introduction of a pathogen to an area from which it has previously been absent can cause major outbreaks of disease in plant communities. The amount of disease that develops is often determined by Pathogencity of the pathogen. Pathogencity is nothing but the inherent ability of the pathogen to cause disease and is dependent on the pathogens, reproductive, dispersal and survival fitness.
2. ***Host:*** The development of disease in a plant community relies on the presence of individual hosts that are susceptible to that particular pathogen. If the majority of the population is susceptible to the pathotypes of a pathogen in the vicinity, an epidemic can occur. The best way of controlling disease is by planting species or cultivars that are not susceptible to pathogen of that area. The occurrence of disease can also be influenced by the host plants growth stage and form same diseases are common in seedling, while others are typical of mature plants. The growth stage of the population can also affect the microclimate around the plants.

3. ***The Environment:*** The presence of a pathogen against a particular plant will generally not cause serious disease unless the environmental conditions are favourable. This includes the aerial environmental and the soil (edaphic) environmental. Human attempts at controlling disease usually involve manipulating the environment in some way. For example, breeding wheat cultivars to tolerate dry conditions allows Australian farmers to plant the crop in areas that are not favourable for pathogens such as powdery mildew and leaf rust. Properties of the aerial environment that influence disease development include moisture levels, temperature and pollution. Moisture is particularly important to pathogenic bacteria and fungi. Rain splash plays an important role in the dispersal of some fungi and nearly all bacteria, and a period of leaf wetness is necessary for the germination of most airborne spores. By using water for dispersal, propagules are dispersed at a time when they are likely to be able to germinate as well. Because the process of germination and infection takes time, the duration of leaf wetness also influences the success of the infection. The duration necessary for infection various with temperature. Usually, a longer period of leaf wetness is needed to establish an infection in cooler temperature, as germination and infection are generally accelerated in warmer conditions. Twemperature also affects the incubation or latent period, the generation time and the infection period. The disease cycle speeds up at higher temperatures resulting in faster development of epidemics.

Methods of Disease Forecasting

Disease forecasting requires field observations on the pathogen characters, collection of weather data, variety of the crop and certain investigations and their correlations. Usually the following methods are employed in disease forecasting.

Forecasting Based on Primary Inoculums

Presence of primary inoculums, its density and viability are determined in the air, soil or planting material. Occurrence of viable spores or propagules in the air can be assessed by using different air trapping devices (spore traps). In the case of soil-borne diseases the primary inoculum in the soil can be determined by monoculture method. Presence of loose smut of wheat, ergot of pearlmillet and viral diseases of potato can be detected in the seed lots at random by different seed testing methods. Seed testing methods can be used to determine potential disease incidence and enable decision to be made on the need for chemical seed treatment. The extent of many virus diseases is dependent on the severity of the preceding winter which affects the size of vector population in the growing season. *e.g.*, Sugarbeet yellows virus.

Three types of diseases can be predicted from the knowledge of initial inoculum or initial disease monocyclic disease, polycyclic disease for which the pathogen has few generation, polycyclic disease for which the initial inoculums is potentially large (Fry, 1987). Indirect assessment of pathogen populations have proved useful in forecasting fire blight of apple and pear (*Erwinia ameylovora*) in California (Thomson *et al.*, 1977) same diseases caused by aerially dispersed polycyclic pathogens are forecast on the basis of pathogen spores in the air are trapped and population size in determined spore trapping has been used to forecast early blight of potato caused by *Alternaria soloni* early in the season most spore are produced from mycelium on plant debris later spores form mycelium in infected tissue contributed to the aerial population (Harrison *et al.*, 1965).

1. Forecasting Based on Weather Conditions

Weather conditions *viz.*, temperature, relative humidity, rainfall, light, wind velocity *etc.*, during the crop season and during the inter crop season are measured. Weather conditions above the crop and at the soil surface are also recorded. Weather conditions during the intercrop period the survival of plant pathogen or its vector depends on the weather conditions during the intercrop period. The intercrop weather which favours over wintering or over-summering of the inoculum can be utilized disease prediction.

Example: The severity of blue mold of tobacco (*Peronospons tobacina*) in southern us is related to temperature in the winter season. When January temperature is above normal, blue mold can be expected to appear early in seed beds in the following season and to course seven losses (Agrios, 1988).

2. Forecasting Based on Correlative Information

Weather data of several years are collected and correlated with the intensity of the diseases. The data are compared and then the forecasting of the disease is done. Forecasting criteria developed from comparisons of disease observation with standard meteorological data have been provided for diseases like *Septoria* leaf blotch of wheat, fire blight of apple and barley powdery mildew.

In late blight of potato and tomato the initial inoculum is usually low and generally too small detect and measure directly even with low initial inoculum the initiation and development of a late blight epidemic can be forecast with resorable accuracy if the moisture and temperature condition in the field remain within celain ranges favourable to the pathogen when consent cool temperature between 10 and 24°C prevail and the relative humidity remain over 75 % for atleast 48 h infection will take place and a late blight outbreak can be expected from 2 to 3 weeks later.

1. Use of Computer for Disease Forecasting

In some advanced countries forecasting of disease is made by the use of computers. This system gives the results quickly. One such computer based programmes in the USA is known as 'Blitecast' for potato late blight.

2. Types of Diseases Forewarning System

Starting from simple rule of thumb, the disease prediction can go to very complex models involving several analytical tools. The purpose of a model is to help grower to arrive at a decision about crop health management. Some of these models also could go as sub- routines in crop growth models which are useful to ultimately work out the predicted yields. However, due to complexity of the processes involved in disease development, very few crop growth models have considered including the disease sub routines. Even such models like DSSAT do not take all parameters of disease development and thereby leaving unaccounted variability. Various techniques of computer modelling and simulation like artificial neural networks, fuzzy logic or Montecarlo analysis and the conventional regression approaches are being used to help synthesize and develop scientists understanding of this complex plant-pathogen-environment relationship. The resultant models enable exploration of the factors that govern disease epidemics.

Thumb rules: Thumb rules are qualitative knowledge-based on past experience of definitive relationships among host, pathogen and environment. Dutch rules for prediction of late blight of potato, Mill's rules for prediction of apple scab are some examples.

1. Empirical or Mechanistic Models

These models are developed based on elaborate experiments and deriving simple relationships between various parameters associated with disease development. These are mainly descriptive models covering various weather parameters and their influence on disease development. However, these models cannot be used for quantification or analysis of disease purposes.

2. Mathematical Models

Mathematical models are development by applying the mathematical relationships between host, pathogen and environment interactions. The analytical models use the epidemic analysis on a theoretical basis without taking into consideration the external factors. These include parameters such as apparent rate of infections, disease development under non restrictive environment, equations developed for fitness and survival of the pathogens *etc.* The statistical models include use of statistical tools such as correlation, regression methods. These models could be linear or non – linear relationships between various interacting parameters of a disease. Trapman (1994) developed the Relative Infection Measure (RIM) model in apple against disease *Venturia inequalis*. The model depends on two sub–models based on ascospore spore maturity, germination and penetration of the fungus into fruit. Based on this model, software called RIMpro has been developed, validated and used in many countries.

The simulation models include the art and science of constructing composite models and mimicking one system with the other. Simulation models, when run on the computer, generate growth parameters by considering algorithms based on a detailed flow diagram of the events of disease development. The first computer-based plant disease simulator is EPIDEM for early blight of potato and tomato. Later several simulation models were developed either as a stand-alone model or as a sub-routine of the crop growth model. With the advent of internet, web-based decision support systems are also in place for various diseases. The web-based models are easy to develop as they do not need expertise of computer programs and the web pages could be written with a relative ease as compared to complicated programs such as Java, J-Builder *etc.* Magarey *et al.* (2005) developed a generic infection model based on temperature and wetness duration. The model is generic in the sense that it was developed to describe any pathosystem when appropriate parameters are supplied.

The production and availability of decision support systems (DSS) as aids in applying or using disease control methods have become common in contemporary agriculture, perhaps because it has become easier to obtain and process the data that are needed to provide this kind of advice. Another factor affecting adoption of DSS is the ability to reach large audiences with relatively little effort. While increased availability of this kind of information is undoubtedly a benefit for most users, the quality of such information should be sufficient to allow maximization of the economic and environmental benefits that can result from the use of such systems. Thus, other ways of development of such systems could be useful. In addition, clear and objective methods are needed to evaluate and compare different DSS, although the ultimate criterion is adoption by the end-users.

With the development of more rapid computing technologies, it is possible to compute a large number of simulations to support new modelling methods for plant disease forecasting. The development of disease forecasting models is based on determining the value of unknown parameters for an appropriate model. One method for estimation of parameters is maximum likelihood estimation (MLE). The basic idea behind MLE is that one can calculate the probability of observing a particular set of data given the model of interest and this is called the likelihood of a sample. The likelihood of observing the actual data for each of the potential parameters values is calculated to determine which set of parameter values is calculated to determine which set of parameter values maximizes the likelihood. One disadvantage with mathematical models is that they can be developed and validated using data generated from controlled experiments. However, while extrapolating to field conditions due diligence must be adapted to avoid erroneous conclusions.

1. Stochastic Models

These models are based on probability distributions and generally adopted when a reliable estimate of the expected value is desired and when there is a curvilinear relationship between stochastic characteristic and one of the state variables. From these models generally variance of the model is expected output. These models include heavy calculations needing electronic calculation aids. These models are more common in human and animal diseases where the target organisms are mobile without a continuum and the probability is involved in such cases.

Examples of well developed forecasting systems are given below.

A. Early and Late Leaf Spots of Groundnut

A technique has been developed for forecasting early and late leaf spots of groundnut in the U.S.A. When the groundnut foliage remains wet for a period greater than or equal to 10 h and the minimum temperature is 21°C or higher for two consecutive days or nights, the disease development is forecasted. A computer programme has been developed in the USA. This is accurate and is widely used in the USA. The data on hours for day with relative humidity (RH) of 95 % and above and minimum temperature (T) during the RH observations for the period, for the previous 5 days are fed to the computer. Calculations are rounded to whole numbers. The T/RH index for each of the five days is calculated *e.g.*, when hours of the RH 95 % equal 10 and the minimum temperature during the period equals 21.1°C the T/RH index is 2.0.The T/RH indices for days 4 and 5 are summed. If the total index exceeds 4 disease is forecasted. If the index is 3 or less no disease is forecasted.

B. Late Blight of Potato

In the USA a forecasting programme has been developed for late blight of potato (*Phytophthora infestans*). The initial appearance of late blight is forecasted 7 to 14 days after the occurrence of 10 consecutive blight favourable days. A day is considered to be blight favourable when the 5 day average temperature is 25.5°C and the total rainfall for the last 10 days is more than 3.0 cm. A computerized version (Blitecast) has also been developed in the U.S.A for forecasting potato late blight. Blitecast is written in Fortran IV. When a farmer desires blight cast (blitecast) he telephones the blight cast operator and reports the most recently recorded environmental data. The operator calls for the blight cast programmes in the computer *viz.*, typewriter terminal and feeds the new data into the computer. Within a fraction of second the computer analyses the data and series of a forecast and spray recommendations to the operator who relays it to the farmer. The entire operation can be completed during standard three minutes telephone call. The system makes one of the four recommendations *viz.*, no spray, late blight warning, 7 days spray schedule or 5 days spray schedule. The last 5 days spray schedule is issued only during severe blight weather. In West Germany, 'Phytoprog' is the programme used. It is based on measurements of temperature, relative humidity and rainfall. Phytoprog provides a negative prognose (an indication of when the usual routine spray application should be dispensed with).

C. Blister Blight of Tea

A system for predicting epidemics of blister blight of tea (*Exobasidium vexans*) has been developed based on the number of spores in the air in the tea plantation and the duration of surface wetness on the leaves. The duration of sunshine is negatively correlated with the duration of surface wetness. The following prediction equation has been developed.

$$Y = 1.8324 + 0.8439\ X1 + 0.9665\ X2 - 0.1031\ X3$$

where, X1 = log % infection t2

X2 = log % infection t^2 – log infection t^1 Y = log of the number of spores in the air and $t^1 - t^2$ three weeks

X3 = mean daily sunshine for a 7 days period preceding t^2

D. Southern Corn Leaf Blight

'Epimay' is a system for forecasting Southern corn leaf blight (*Bipolaris maydis*) based on conceptual model.

A. Rice Blast

In India, forecasting rice blast (*Pyricularia oryzae*) is done by correlative information method. It is predicted on the basis of minimum night temperature 20 to 26°C in association with high relative humidity of 90 % or above. Computer based forecasting system has also been developed for rice blast in India.

B. Wheat Stem Rust

Forecasting wheat stem rust epidemic is done by analysing the rain samples which give precise data for inoculum present in the air. Moreover several wind projectors are also prepared to survey the air-borne primary inoculums and its deposition. It has been observed that primary inoculum comes from South India, to the plains of Central and North India.

C. Brown Stripe Downy Mildew of Corn

The forecasting of brown stripe downy mildew of corn (*Sclerophthora raysise* var. *zeae)* which is restricted to India is done on the basis of average rainfall 100 to 200 cm or more accompanied by low temperature (25°C or less). Spore trapping Techniques of acquisition of biological data for consecutive forecasting models are important. Spore traps have been widely used in to complete disease with weather conditions. Spore trapping is useful for understanding epidemiology of a disease and behaviour of the pathogens. This helps in developing models on dispersal of pathogens or on epidemiology of the disease and to formulate methods of management. Methodology of spore trapping depends on the following objectives of the worker.

1. Biology of the pathogen
2. For infection forecasting
3. Spore dispersal gradients
4. Management of the disease

In epidemics of air-borne plant diseases the number of spores of the pathogen landing on the plant which depends on the number of spores in the atmosphere above the crop is an important factor for the quantitative sampling of the atmosphere (number of spores per unit volume of air). For trapping and estimating these studies different types of traps are used. The following spore traps are usually employed in trapping of fungal spores. Cylindrical rods or microscopic glass slides: It helps to gather data on the spore arrival in a locality. In this, the surface of microscopic slide is smeared with grease and made sticky. In the method, quantitative estimation is not possible as number of spores collected is very low.

1. Hirst's Volumetric Spore Trap (Hirst, 1952)

In this instrument, air is sucked into at a controlled rate and impinged on to a glass slide moved by a clockwork mechanism past the orifice. It gives continuous count of spores in 24 h. The number of spores per unit volume of air at any given time can thus be calculated.

2. Rotorod Sampler or Rotorod Spore Trap (Sutton and Jones, 1976)

It comprises of a 'U' shaped rod attached at its mid point to the shaft of a small battery operated electric motor. In this equipment the surface of the rod is covered with a vaseline strip of transparent cellophanes to catch spores which can be taken off and mounted on a glass slide. From the area of the strip and the speed of rotation, the volume of air samples can be calculated.

3. Anderson Cascade Spore Sampler

It is a device where Petri plates with nutrient agar are used to collect the spores.

4. Bourdillon Slit Sampler

Air is sucked in a chamber by vacuum pump which strikes the rotating Petri dish containing agar medium. The agar medium retains the spores sucked in the air. Concentration of viable spores is calculated after counting germinated spores in the medium.

5. Burkard's 7 Day Volumetric Spore Trap

This device records spores in the air drawn by a pump on 7 days basis on a cellophane strip wrapped on a drum rotating inside a chamber.

6. Jet Spore Trap

In the above sampling methods, the viability of the spores cannot be determined. To overcome this, living plants have been used as spore traps. A jet spore trap in which spores are impacted in an air jet into a column of still air, through which they fall, to settle on leaf segments exposed at the base of the chamber. In this trap, suitable cultivars of host plants can be employed to determine number of viable spores.

References

Agrios, G. N., 1988, Plant Pathology, 3rd ed., Academic Press, New York, 803.

Bourke, P. M. A., 1937, Potato blight and the weather in Ireland (Tech. Note No. 15) in 1953, Irish Meteorol. Serv., Dublin, 15.

Cowling, E.R., and Horsfall, J.G., 1978, Prologue: how disease develops in populations, in Plant Diseases: An Advanced Treatise, Vol. 2, Horsfall, J. G. and Cowling, E. B., Eds., Academic press New York, 436.

Fry, W. E., 1987, Principles of Plant Disease Management, Academic Press, New York, 378.

Harrison, M.D., Livingston, C.H. and Oshima, N., 1965, Control of potato early blight in Colorado, Fungicidal spray schedules in relation to the epidemiology of the disease., Am. Potato. J., 42: 319.

Hriday, S.Chaube and Uma, S, Singh, 1990, Plant disease Management : Principles and Practions, CRC, press., 319.

Hyre, R. A. Bonde, R., and Johnson, B., 1959, The relation of rainfall, relative humidity, and temperature to late blight in Maine, Plant Dis. Rep., 43, 51.

Hyre, R. A. Bonde, R., and Manzer, F. E., 1960, Re-evaluation in Maine of three methods for forecasting late blight of potato, Plant Dis. Rep., 44, 235.

Magarey, R.D., Sutton, T.B. and Thayer, C.L., 2005, A simple generic infection model for foliar fungal plant pathogen. Phytopathology, 95: 92-100.

Nagarajan, S and Singh, H., 1975, Indian stem rust rules-an epidemiological concept on the spread of stem rust of wheat. Plant Disease Reporter, 59: 133-136.

Pradhan, S., 1967, Strategy of insect pest control. Indian J. Entomology. 29 : 105 -122.

Pradhan, S., 1982, Effect of temperature on the dynamics of insect development.

Agricultural Entomology and pest control. ICAR. New Delhi.

Shrum, R. D., 1978, Forecasting of Epidemics, in Plant Diseases: An Advanced Treatise, Vol. 2, Horsfall, J. G. and Cowling, E. B., Eds., Academic press New York, 436.

Srinivasan, G.S and Mavi, H.S., 1983, Proc. Int. Conf. on Biometeorology, New Delhi.

Thomson, S. V., Schroth, M. N., Reil, W. O., Beutel, J. A. And Davis, C. S., 1977 Pesticide applications can be reduced by forecasting the occurrence of fireblight bacteria, Calif. Agric., 31, 12.

Van der Plank, J.E., 1963, Plant disease Epidemic and Control, Acdemic Press, New York, 347.

Van Everdingen, E.1926, Het verband de wears gesterdheid ende aardappelzeiekte (*Phytophthora infestans*). Tijdschr. Plantennziekten. 32: 129-140.

11

Indian Monsoon and Challenges In Monsoon Forecasting: IMD's Future Plans

Vishwanath Biradar

Introduction

Due to its unique geo-climatic conditions Indian subcontinent experiences a variety of weather systems like monsoons, tropical cyclones, thunderstorms, snow storms *etc.* The region is also affected by a number of natural hazards like tropical cyclones, floods, droughts, earthquakes, landslides, avalanches *etc.* Past records show that 68 % of the area is susceptible to drought and 40 million hectares prone to floods.

The most important weather system from Indian point of view undoubtedly is the southwest monsoon. India, primarily an agriculture country receives about 75 % of its annual rainfall due to southwest monsoon and hence, prediction of various phases of monsoon, like its onset & advance, active and break cycle *etc.* is very crucial to the farming community. Prediction of rain producing monsoon systems like monsoon depressions and lows and the monsoon rainfall pattern in different time scales is important for many sectors like agriculture, water management, power industry *etc.* Most of the major floods also occur in this season due to heavy rains and for flood management it is very important to predict heavy rainfall events well in advance. A failure in monsoon rainfall leads to drought conditions and affect the economy of the country. Monsoon rainfall prediction in long range is very useful for planners, especially when a large deficient rainfall scenario is foreseen.

Thus, monsoon throws many serious problems to the operational as well as research community. Being the National Meteorological Services agency, India Meteorological Department (IMD) has the responsibility to monitor and analyze the monsoon activity and provide weather forecasts/warnings in different time scales for various sectors like agriculture, water resources, power sector and irrigation department as well as general public and disaster managers.

Monsoon Forecasting – Important Issues

Climatologically monsoon season is from June to September and the country receives about 75 % of its annual rainfall in this season (Rao, 1976). Generally monsoon sets in Kerala on 1st June and covers entire country by 15th July. An abnormal delay in the onset or advance of monsoon may lead to little rainfall over the region. Month-wise, July and August contribute maximum to the seasonal rainfall. Monsoon withdrawal generally commences in the first week of September from extreme northwest India and by middle October, it withdraws completely from India. A few monsoon depressions and low pressure systems form over northwest Bay of Bengal in this season and cause heavy rainfall as they traverse westwards leading to floods and sometimes damaging the standing crops. Whenever monsoon through shifts northward towards to the foothills of Himalayas and sticks there for some days, it may give rise to break monsoon situation with marked decrease in rainfall over major parts of the country, but increased rainfall all along foothills. A prolonged break condition not only severely affects the agriculture sector, but also leads to All India drought situation.

(A) Seasonal Prediction of Monsoon Rainfall

IMD issues operational Long Range Forecasts (LRF) for the southwest monsoon rainfall at two stages. First stage forecast is issued in April and the second stage forecast is issued in June. Since 2007, IMD is using the statistical models for forecasting the monsoon rainfall (June-September) for the country as a whole.

(B) Prediction of Active/Break Monsoon Spell

It is observed that monsoon sometime exhibits active and break/weak spells of rainfall activity within the season for example in the year 2002, whole July showed large deficient rainfall followed by nearly good August. Prediction of such weak and good spells of rainfall activity is very important for agriculture sector and water management but very challenging. Both statistical models and dynamical models have only limited success in these aspects. In recent years some empirical techniques (for example: Webster and Hoyos (BAMS) (wavelet banding) Goswami *et al.*,/Xavier and Goswami. (2007) have shown some good success in predicting active/break spell. However more research efforts are ongoing to identify better forecasting tools for forecasting active/break monsoon spells. Regional climate models also showing potential this regard.

(C) Prediction of Extreme Rainfall Events

During monsoon season there are occasion when heavy rainfall may occur in a short spell causing flash flood and damage to the property/crop. The problem becomes very serious if such extremely heavy rainfall occurs over metro cities, for example Mumbai rainfall (26-27 July, 2005). Prediction of such heavy events atleast in short range duration is very important for disaster managers and general public. However, the present forecasting system has limitations in forecasting such extreme events and efforts are in progress to tackle such issues by enhancing the observation, developing more sophisticated numerical models with accurate data assimilation techniques *etc.* Definitely this topic needs more concerted efforts by all the research groups to further help operational forecasters.

IMD's Future Plan

IMD has taken up major initiatives for providing more efficient weather services by strengthening and upgrading its observational, communication and forecasting system. The salient features are given below:

1. To establish a nationwide network of AWS, high density satellite based automatic raingauge stations, wind profilers and lightning detection systems *etc.*
2. To introduce advanced numerical models
3. To acquire high performance computing system
4. Generation of drought outlook
5. Generation of extended range forecast (11-30 days) using both statistical and dynamical models

Some of the modern instruments are already in place and assisting the operational forecasters. Many products and forecasts are put on the IMD's official website at Head Quarters as well as regional websites. IMD has taken up steps to introduce National weather channels for providing more frequently the weather related information and forecasts/warnings to various users including general public. It is also proposing to create district meteorological centers for quick and improved meteorological services to the farmers and other users. Hopefully, with induction of advanced observing systems, high resolution models and modern communication facilities, IMD's overall capabilities will be strengthened and the ultimate goal of providing more accurate and improved weather and climate services to the nation will be fulfilled.

12

Role of Green Nanoparticles in the Management of Emerging Plant Diseases in Relation to Climate Change

V.B.Nargund, Chikkanna Swamy,
K.N. Basavesha and H.N. Srikanth

Introduction

Applications of nanotechnology in food and agricultural sciences are exciting to consider but implications involve risks for both the environment and organisms including human beings. Agricultural scientists are having a wide spectrum of challenges such as lower agricultural input efficiency, delivery of nutrients and pesticides, unsustainable farm management practices, water and vulnerabilities to climate change due to global warming.To address these research activities, there is a need to explore one of the frontier technologies such as 'Nanotechnology' to precisely detect and deliver the correct quantity of nutrients and pesticides that promote productivity while ensuring environmental safety and higher use efficiency. The nanotechnology can be exploited in the value chain of entire agriculture production system (Subramanian and Tarafdar, 2011). Nano-phytopathology is a cutting-edge science which uses nanotechnology for detecting, diagnosing and managing plant diseases and their pathogens at an early stage. Nanodiagnosis by portable polymerase chain reaction (PCR) systems, nanopore sequencing tools, nanodiagnostic kit, gold nanoparticles, quantum dots (QDs), nanobarcodes and nano-sensors are few examples. Modern nanomolecular techniques are used for monitoring or understanding of pathogen population genetics, plant microbe interactions and gene transfer between pathogens and even the host. Additionally, nanomaterials can be used for mycotoxin detection and detoxication, increasing plant resistance, plant disease forecasting and nano-molecular diagnostics of plant pathogens (Khiyami *et al.*, 2014).

The integration of the principles of green chemistry to nanotechnology towards the synthesis of green nanoparticles is a current requirement in plant pathology. Green principle route of synthesizing nanoparticles have emerged as alternative to overcome the limitation of conventional methods among which plant and microorganisms are majorly exploited.

Applications of Green Nanoparticles in Plant Diseases Management

Crop plants are affected by newly emerging and re-emerging pathogens due to change in climatic conditions and also presently many pathogens are getting resistance to available fungicides. Climate change is one of the potential factors for adaptation and development of virulent races in plant pathogens. In the present day it is necessary to detect early the plant pathogens both in field and quarantine station by novel technological approaches. So, management of plant diseases is a challenging task for maintaining the sustainable production. Diseases are one of the major limiting factors in crop productivity. The problem with disease management lies with the detection at exact stage of prevention but most of times pesticides are applied as a precautionary manner leading to residual toxicity and environmental hazards. On the other hand application of pesticides after the appearance of disease leads to some amount of crop losses. Use of micronized chemical pesticides in controlling plant diseases is an age old practice. In its comparison, nanoformulation is a recent introduction to the field of plant protection. The efficiency can be improved by generating nanoparticles (NPs) and hence nanotechnology attains much attention. Hence scientists are always searching for new management strategy. Now Green Nanotechnology is a novel and emerging research areas in agriculture and allied sciences including plant pathology. Synthesis of nanoparticles from plant/microbes is more fascinating and eco friendly compared to chemical methods in biological sciences including agricultural science.

1. Nanosized metals as diagnostic probes: Biosensor/nanosensors
2. 'Smart' delivery- Nanomaterials for management of plant diseases

Nanosized silver, Nanosized silica-silver, Silica nanoparticles, Nano-copper, Nano-iron and Carbon nanotubes are few examples presently used in plant disease management.

Nanosized Metals as Diagnostic Probes for Detecting Plant Diseases

A wide variety of sensors have been developed and commercialized for various applications including environmental monitoring and medical diagnostics. Depending on the operating principle of the sensor, the analytes could be detected using a sensor based on electrical, chemical, electrochemical, optical, magnetic or vibrational signals. The limit of detection could be enhanced by the use of nanomaterial matrices as transducers and the specificity could be enhanced by the use of bio-recognition elements such as DNA, antibody, enzymes *etc.*

A need for detecting plant disease at an early stage so that tons of food can be protected from the possible outbreak; has tempted nanotechnologists to look for a nano solution for protecting the food and agricultural product from bacteria, fungus and viral agents. A detection technique that takes less time and that can give results within a few hours, that are simple, portable and accurate and does not require any complicated technique for operation so that even a simple farmer can use the portable system. If an autonomous nano-sensors linked into a GPS system for real-time monitoring can be distributed throughout the field to monitor soil conditions and crop, it would be of great help. The union of biotechnology and nanotechnology in sensors will create equipment of increased sensitivity, allowing an earlier response to environmental changes and diseases.

Nanoparticles have been utilized with other biological materials such as antibody for detecting *Xanthomonas axonopodis* that causes bacterial spot disease (Yao *et al.*, 2009). Gold nanoparticle-based optical immunosensors have been developed for detection of karnal bunt disease in wheat using surface plasmon resonance (SPR) (Singh *et al.*, 2010). Fluorescent silica nanoparticles (FSNPs) combined with antibody as a biomarker have been studied as the probe, which successfully detected plant pathogens such as *Xanthomonas axonopodis* pv. *vesicatoria* that cause bacterial spot disease in solanaceae plant. Many QD-FRET (Quantum Dots Fluorescence Resonance Energy Transfer) based sensors have been developed for phytodisease detection such as the Witches' Broom Disease of Lime (WBDL) caused by Candidatus *Phytoplasma aurantifolia* (Ca. *P. aurantifolia*). The immunosensor constructed showed a high sensitivity and specificity of 100%, and a detection limit of 5 ca. *P. aurantifolia*/µl (Rad *et al.*, 2012).

Additionally, QD-FRET was also reported to detect the disease vectors. For example, Rhizomania, which is the most destructive disease in sugar beet, is caused by Beet necrotic yellow vein virus (BNYVV). *Polymyxa betae* (Keskin), the only known vector of BNYVV, for transmission of the virus to the plants was successfully reported to be detected by QD-FRET-based sensor (Safarpour *et al.*, 2010). In recent years, antibody-based biosensor technology has seen tremendous progress upon implementation of nanotechnology based approaches for the sensor fabrication. Gold nanorods (AuNRs) functionalized by antibodies have been used to detect Cymbidium mosaic virus (CymMV) or Odontoglossum ringspot virus (ORSV) for rapid diagnosis of viral infections (Lin *et al.*, 2014). A novel portable cell biosensor system for detection of Potato virus Y (PVY), Cucumber mosaic virus (CMV) and Tobacco rattle virus (TRV) was fabricated by immobilizing the vero cells carrying virus specific antibodies on their membranes. This study demonstrated an important step towards the development of a portable plant virus detection system suitable for on-field application (Kumar *et al.*, 2015).

Nanoparticles Based Smart Delivery Systems

Smart delivery systems have the capacity to monitor the effects of the delivery of pharmaceuticals, nutraceutical nutrients, food supplements, bioactive compounds, probiotics, chemicals, insecticides, fungicides, vaccinations. In order to protect the active ingredient from the adverse environmental conditions and to promote persistence, a nanotechnology approach, namely "nano-encapsulation" can be used to improve the insecticidal value. Nanoencapsulation comprises nano-sized particles of the active ingredients being sealed by a thin-walled sac or shell (protective coating). Nanoencapsulation of insecticides, fungicides or nematicides will help in producing a formulation which offers effective control of pests while preventing accumulation of residues in soil.

In order to protect the active ingredient from degradation and to increase persistence, a nanotechnology approach of "controlled release of the active ingredient" may be used to improve effectiveness of the formulation that may greatly decrease amount of pesticide input and associated environmental hazards. Nano-pesticides will reduce the rate of application because the quantity of product actually being effective is reduced least 10-15 times that applied with classical formulations, hence a much smaller than the normal amount could be required to have much better and prolonged management. Several pesticide manufacturers are developing pesticides encapsulated in nanoparticles (Anon., 2008). These pesticides may be time released or released upon the occurrence of an environmental trigger (for example, temperature, and humidity, light). It is unclear whether these pesticide products will be commercially available in the short term. Clay nanotubes (halloysite) have been developed as carriers of pesticides at low cost, for extended release and better contact with plants, and they will reduce the amount of pesticides by 70-80 %, thereby reducing the cost of pesticide with minimum impact on water streams.

Synthesis of Green Nanoparticles

There are number of means to produce metallic nanoparticles (NPs) that includes physical approaches like 'attrition' and 'pyrolysis', chemical approaches like wet chemical methods. Although these procedures are widely used but they have got some drawbacks like enormous consumption of energy in order to maintain high pressure and temperature in physical methods, use of toxic solvents and generation of hazardous biproducts in chemical methods. Hence, cost effective, non-toxic and environmentally benign procedures for the synthesis of metallic nanoparticles are being used for which biological approaches have gained increasing importance (Thakkar *et al.*, 2010). Biological resources such as plants and plant products, algae, fungi, yeast and bacteria could be employed for the synthesis of nanoparticles. Metal nanoparticles synthesized with different plant parts and applications were used in various microbial activities (Haleemkhan *et al.*, 2015). Plants and microbes are the bio-factories for many phyto-chemicals and reducing agents. These phyto-chemicals are responsible for reduction of bulk material to nanoform. Silver, copper, zinc and sulphur nanoparticles are synthesized from various plant spices and plant parts like leaves, flowers, fruits and stems further microbial antagonist (fluorescent *Pseudomonas*) and chitosan from shrimps were also used. Various procedure starting from High energy ball mill, heating (muffle furnace, microwave and magnetic stirrer) and sunlight exposure with aqueous extract of plants and chitosan were attempted (Chikkanna Swamy and Nargund, 2017; Nargund *et al.*, 2016a: 2017a).

Effect of Nanoparticles on Plant Pathogens

Silver nanoparticles have strong potential as an antimycotic activity against fungi of grey mold, *Botrytis cinerea* (Oh *et al.*, 2006). Park *et al.* (2006) also studied the 'effective concentration' of nanosized silica silver on suppression of growth of many fungi; and found that, *Pythium ultimum, Magnaporthe grisea, Colletotrichum gloeosporioides, Botrytis cinerea and, Rhizoctonia solani*, showed 100% growth inhibition at 10 ppm of the nanosized silica-silver. Whereas, *Bacillus subtilis, Azotobacter chrococuum, Rhizobium tropici, Pseudomonas syringae* and *Xanthomonas compestris* pv. *vesicatoria* showed 100% growth inhibition at 100 ppm Kirby-Bauer disc diffusion protocol was used to evaluate the communal effect of fluconazole and silver nanoparticles for their antifungal activity against three fungal pathogens, *Phoma glomerata, Phoma herbarum*, and *Fusarium semitectum* (Gajbhiye *et al.*, 2009).

Ag_2S nanocrystals on amorphous silica particles show antifungal activity against *Aspergilus niger*. The potential biocidal efficacy of ZnO and $ZnTiO_3$ nanopowders against the fungus *A. niger* was assessed (Ruffolo *et al.*, 2010). $ZnTiO_3$ nanopowder showed higher growth inhibition efficiency than ZnO. The antifungal activity of the silver nanoparticles was evaluated on the phytopathogen *Colletotrichum gloeosporioides*, which is responsible for anthracnose in a wide range of fruit crops. Silver nanoparticles significantly reduced the mycelia growth of *C. gloeosporioides* in a dose dependent manner (Mendez *et al.*, 2011). Antifungal properties of silver nanoparticles, silver ions, acrylate paint and cotton fabric impregnated with silver nanoparticles were assessed against *A. niger, Aureobasidium pullulans* and *Penicillium phoeniceum* (Khaydarov *et al.*, 2011). Bioassay of elemental and nanosulphur against *A. niger* showed that nanosulfur was more efficient than its elemental structure (Choudhury *et al.*, 2010).

Antifungal activity of nanoparticles of zinc oxide (35-45 nm), silver (20-80 nm) and titanium dioxide (85-100 nm) has been tested against *Macrophomina phaseolina*. The higher antifungal effect was observed in silver nanoparticles at lower concentrations than zinc oxide and titanium dioxide nanoparticles (Shyla *et al.*, 2014). Maize treated with nanosilica (20-40 nm) has been screened for resistance against phytopathogens, *Fusarium oxysporum* and *Aspergillus niger* as compared with that of bulk silica. Nanosilica-treated plant showed a higher expression of phenolic compounds (2056 and 743 mg/ml respectively) in collected leaf extracts and a low expression of stress-responsive enzymes against these fungi. These results showed significantly higher resistance in maize in treated with nanosilica than with bulk in terms of disease index and expression of total phenols, phenylalanine ammonia lyase, peroxidase and polyphenol oxidase, at 10 and 15 kg/ha (Suriyaprabha *et al.*, 2014).

Targeted management of plant disease with lower dose of nano formulations is newly emphasizing analogous to target delivery in medical sciences. Silver, copper, sulphur, zinc nanoparticles were synthesized by plant/chitosan mediated green nanotechnology. The sizes of nanoparticles were confirmed by particle size analyzer and atomic force microscope. Confirmed nanoparticles were tested against plant pathogens *viz.*, *Xanthomonas* spp. *Ralstonia solanacearum, Sclerotium rolfsii, Rhoztonia solani* (Nargund *et al.*, 2016a & b) Synthesized the silver, copper and sulphur nanoparticles from the various plant and plant parts (cotton stems, eucalyptus leaves and pomegranate pericarp) aqueous extract changes in colour from bulk to nanoform is one of the preliminary indications of confirmation of nanoparticles. Further nanoparticles were confirmed by Uv-Visible Spectrophotometer, Particles Size Analyzer, Atomic Force Microscope and Scanning Electron Microscope with EDS with the size of < 60 nm. Green synthesized silver, copper and sulphur nanoparticles were tested against major fungal (*Exserohilum turcicum, Ustilago nuda tritici* and *Golovinomyces cichoracearum*) and bacterial (*Xanthomonas campestris* pv. *campestris* and *X. axonopodis pv. malvacearum*) plant pathogens. Phytotoxicity study was attempted on cowpea, chilli and tomato seedlings. Results showed green nanoparticles have no phytotoxic effects on tested seedlings (Nargund *et al.*, 2017a and b).

Conclusions

Perusal of the literature and critical analysis of the relevant information have related nanotechnology has potential prospects of use and application in the detection, diagnosis and management of plant diseases. The limited studies so far conducted on this aspect are sufficient enough to warrant potential future use of nanotechnology in plant disease management. The direct applications of nanoparticles (Silver, Silica, Copper, Zinc) significantly suppressed the plant pathogens. Environmental fluctuations are much more affected by agriculture and minor diseases became major and cause epidemic some time. So, enzyme based biosensors coated with nanoparticles of Au, Ag, Co, Ti, Cu, Zn *etc.* may greatly help in the precise and quick diagnosis of plant infection in early stages of plant and residue detection of pesticides. Despite of tremendous application scope of nanotechnology in plant diseases management, there are certain, risk and apprehensions in the use of nanoparticles which are required to be worked out on priority before the commercial use of the products.

References

Anonymus, 2008, Sizes that matter: Opportunities and risks of nanotechnologies. Report in cooperation with the OECD International Futures Programme. http:/www.oecd.org/dataoecd/32/1/44108334.pdf.

Chikkanna Swamy and Nargund, V. B., 2017, Sunlight induced mediated silver nanoparticles from seeds of *Thevetia peruviana* L., characterization and their antifungal efficacy against *Curvularia lunata* (Wakker) Boedijn. Int. J. Curr. Microbiol. Appl. Sci., 6(11): 1008-1013.

Choudhury, S. R., Nair, K. K., Kumar R, Gogoi, R. and Srivastava, C., 2010, Nanosulfur: A potent fungicide against food pathogen, *Aspergillus niger*. AIP Conf. Proc. 1276: 154-157.

Gajbhiye, M., Kesharwani, J., Ingle, A., Gade, A., Rai, M., 2009, Fungus-mediated synthesis of silver nanoparticles and their activity against pathogenic fungi in combination with fluconazole. Nanomed., 5: 382-386.

Haleemkhan, A. A., Naseem and Vidya, V., 2015, Synthesis of nanoparticles from plant extracts *Int. J. Mod. Chem. Appl. Sci.*, 2(3): 195-203.

Khaydarov, R. R., Khaydarov, R, A., Evgrafova, S. and Estrin, Y., 2011, Using silver nanoparticles as an antimicrobial agent. *Chem. Biol.*, 28: 169-177.

Kumar, A., Dash, S. K. and Suman, D., 2015, DNA based biosensors for detection of pathogens. In *Plant Fungal Disease Management*, 1st ed.; New York, USA. pp. 31–35.

Lin, H. Y., Huang, C. H., Lu, S. H., Kuo, I. T. and Chau, L. K., 2014, Direct detection of orchid viruses using nanorod-based fiber optic particle plasmon resonance immunosensor. *Biosens. Bioelectron.*, 51: 371-378.

Mendez, M. A., Martin, E. S., Ortega, L., Cobian, G., Sanchez, E., 2011, Synthesis and characterization of silver nanoparticles: effect on phytopathogen *Colletotrichum gloesporioides*. J. Nanopart. Res.13: 2525-2532.

Nargund, V. B. Patil, R. R., Hasansab A. Nadaf, Puneeth Raj M. S., Chikkanna Swamy, Madhu S. Giri, Nagabushan, H., Taranath, T. C. and Bheemanagouda Patil, 2016b, Potential Applications of green nontechnology in crop protection. International Conference on Nanotechnology (ICNano-2016), Visvesvaraya Technology University, Chikkaallapura, Karnataka, p.23.

Nargund, V. B., Chikkanna Swamy, Gulabnabi, L. V., Basavesha, K. N., Rajinkanth, V., Savitha, C., Vinay, J. U., Chidanadappa and Srikanth, H. N. 2017a. Nanotechnology novel tool for diagnosis and management of plant diseases. National Symposium on Emerging trends in plant health management in relation to climate change. College of Horticulture, Bengaluru 2017. p.134.

Nargund, V. B., Chikkanna Swamy, Madhu S. Giri, Hulagappa and Ramanagouda, H., 2016a, Green nanotechnology: Synthesis, characterization of silver and copper nanoparticles and their efficacy against major pathogens of maize and pomegranate. *Indian Phytopath.*, 69 (4s): 598-601.

Nargund, V. B., Chikkanna Swamy, Vinay, J.U., Patil, R. R., Byadgi, A. S., and Nadaf, H.A., 2017b, Eco friendly management of major plant pathogens by green nanotechnology: An overview. National Symposium on Diagnosis Management of plant Disease: Integrated Approaches and Recent Trends. Umiam, Meghalaya 2017. p.123.

Oh, S. D., Lee, S., Choi, S. H., Lee and Lee, Y. M., 2006, Synthesis of Ag and Ag-SiO_2 nanoparticles by y-irradiation and their antibacterial and antifungal efficiency against *Salmonella enteric* and *Botrytis cinerea*. *Colloids and Surfaces A: Physicochem Eng Aspects*. 275: 228-233.

Park, Sung, Hwa and Seong, C., 2006, A new composition of nanosized silica-silver for control of various plant diseases. *Plant Pathol. J.* 22(3): 295-302.

Rad, F., Mohsenifar, A., Tabatabaei, M., Safarnejad, M. R., Shahryari, F., Safarpour, H., Foroutan, A., Mardi, M., Davoudi, D. And Fotokian, M., 2012, Detection of Candidatus *Phytoplasma aurantifolia* with a quantum dots fret-based biosensor. J. Pl. Pathol., 94, 525-534.

Ruffolo, S. A., Russa, M. F., Malagodi, M., Oliviero, Rossi, C. and Palermo, A. M., 2010, ZnO and $ZnTiO_3$ nanopowders for antimicrobial stone coating. *Appl. Phys.*, 100: 829-834.

Safarpour, H., Safarnejad, M. R., Tabatabaei, M., Mohsenifar, A., Rad, F., Basirat, M., Shahryari, F. and Hasanzadeh, F., 2012, Development of a quantum dots FRET-based biosensor for efficient detection of *Polymyxa betae*. *Can. J. Plant Pathol.*, 34, 507-515.

Shyla, K. K., Natarajan, N. and Nakkeeran, S., 2014 Antifungal activity of zinc oxide, silver and titanium dioxide nanoparticles against *Macrophomina phaseolina*, *J. Mycol. Plant Pathol.*, 44: 268-273.

Singh, S., Singh, M., Agrawal, V. V. and Kumar, A., 2010, An attempt to develop surface plasmon resonance based immunosensor for Karnal bunt (*Tilletia indica*) diagnosis based on the experience of nano-gold based lateral flow immuno-dipstick test. *Thin Solid Films*, 519: 1156-1159.

Suriyaprabha, R., Karunakaran, G. K., Kavitha, R., Yuvakkumar, V., Rajendran, N., Kannan, Application of silica nanoparticles in maize to enhance fungal resistance, *Nanobiotechnol.*, 8: 133-137.

Thakkar, Kaushik N., Mhatre, Snehit S. and Parikh, Rasesh Y., 2010, Biological synthesis of metallic nanoparticles. Nanomedicine: *Nanotechnol. Biol. Med.*, 6: 257-262.

Yao, K. S., Li, S. J., Tzeng, K. C.; Cheng, T. C., Chang, C. Y., Chiu, C. Y., Liao, C. Y., Hsu, J. J., Lin, Z. P., Fluorescence silica nanoprobe as a biomarker for rapid detection of plant pathogens. *Adv. Mater. Res.* 79: 513-516.

13

Effects of Climate Change on Insect-Host Plant-Environment Interactions

Jagdish Jaba

Introduction

To end of hunger and malnutrition, achieve food security and improve nutrition is at the heart of sustainable development goals. At the same time, climate change is already impacting agriculture and food security, and will make the challenge of ending hunger and malnutrition even more difficult. The effects of climate change on our ecosystems are quite severe and widespread, and ensuring food security in the face of climate change is one of the most daunting challenges facing humankind. While some of the problems associated with climate change are emerging gradually, action is urgently needed to build resilience in agricultural production systems. Despite tremendous progress in crop production, almost 800 million people are chronically undernourished, and 161 million under-five year olds are stunted (FAO, 2016).

According to the IPCC's Fifth Assessment Report (2016), climate change over the last 30 years has already reduced global agricultural production by 1 – 5 % per decade. Recent studies have indicated that a 2 degrees increase in global temperature will affect agricultural productivity, particularly in the tropical regions (Kirtman *et al.*, 2013; Dinesh *et al.*, 2015). Rise in temperature is largely due to increased concentration of greenhouse gases, which include carbon dioxide (CO_2), methane (CH_4), nitrous oxide (N_2O) and chlorofluoro-carbons (CFCs). Over the past 200 years, the atmospheric concentration of CO_2 increased by 35 %, and is expected to double by the end of this century (Houghton *et al.*, 1995; NOAA 2016).

Long-term monitoring of pest population levels and behavior, particularly in identifiably sensitive regions, may provide some of the first indications of a biological response to climate change. It will be important to keep ahead of undesirable pest adaptations, and consider global warming and climate change for planning research and development efforts for integrated pest management (IPM) in the future. The temperature increases associated with climatic changes could result in:

- ✰ Change in geographical range of insect pests,
- ✰ Increased overwintering and rapid population growth,
- ✰ Changes in insect–host plant–natural enemy interactions,
- ✰ Impact on arthropod diversity and extinction of species,
- ✰ Changes in synchrony between insect pests and their crop hosts,
- ✰ Introduction of alternative hosts as green bridges,
- ✰ Changes in relative abundance and effectiveness of biocontrol agents,
- ✰ Change in expression of resistance to insects in cultivars with temperature-sensitive genes,
- ✰ Emergence of new pest problems and increased risk of invasion by migrant pests, and
- ✰ Reduced efficacy of different components of insect-pest management.

Climate Change, Pest Associated Losses and Food Security

To understand how best to control insect pests and diseases to improve food security in the context of climate change, plant protection professionals must work with societal change, defining its key processes and influencers to effect change. Insect pests and diseases could potentially deprive humanity of up to 82 % of the attainable yield in the case of cotton, and over 50 % for other major crops (Oerke, 2006). Insects pests, one of the major constraint in food production, are highly sensitive to climate change, which cause an estimated annual loss of 10-16 % globally (Chakraborty and Newton, 2011), and the pest associated losses in India have been estimated to be 17.5 % (Dhaliwal *et al.*, 2010) valued at US $220 billion. Insect pest associated losses are likely to increase as a result of changes in crop diversity, cropping patterns, cropping intensity and climate change (Table 1). Pest associated losses have increased from an average of 7.2 % during the pre-green revolution period to 23.3 % during the post green revolution period in different crops in India (Dhaliwal *et al.*, 2010). High mobility and rapid population growth will increase the extent of losses due to insect pests.

Table 1: Pest Associated Losses in different Crops during the Pre- and Post-green Revolution in India

Crop	*Pre-green Revolution (Early 1960s)*	*Post-green Revolution (early 2000s)*	*Changes in Pest Associated Losses*
Cotton	18.0	50.0	+ 32.0
Groundnut	5.0	15.0	+ 10.0
Other oilseeds	5.0	25.0	+ 20.0
Pulses	5.0	15.0	+ 10.0
Rice	10.0	25.0	+ 15.0

Crop	*Pre-green Revolution (Early 1960s)*	*Post-green Revolution (early 2000s)*	*Changes in Pest Associated Losses*
Maize	5.0	25.0	+ 20.0
Sorghum& millets	3.5	30.0	+ 26.5
Wheat	3.0	5.0	+ 2.0
Sugarcane	10.0	20.0	+ 10.0
Average	7.2	23.3	+ 16.1

Source: Dhaliwal *et al.* (2010).

Climate Change and Emerging Pest Problems in India

Several insect pests, that were important in the past or are minor pests, are likely to become more devastating with global warming and climate change (Sharma, 2013) (Figure 1 and Table 2). Many insect species, that will move to newer areas as invasive pests will also pose a major threat to crop production and food security, as they find more suitable climatic niches in the new areas. The invasive species are likely to cause more harm in the absence of natural enemies in the new habitats.

Figure 1: Pest Outbreaks Due to cClimate Change (a = Pod borer, *Helicoverpa armigera* damage in pigeonpea following wet weather conditions in Sept – Oct, b = Mealy bug, *Ceroplastodes cajaninae* infestation in pigeonpea under prolonged hot and dry conditions, c = Beet armyworm, *Spodoptera exigua* damage in chickpea triggered by winter rains on Oct – Nov, and d = Pink stem borer, *Sesamia inferens* damage in sorghum due to hot and dry conditions during the postrainy season).

Table 2: Insect Pests that have become or are likely to become Serious Pests Due to Climate Change and Changes in Cropping Patterns

Insect Pest	*Scientific Name*	*Crop(s)*
American bollworm	*Helicoverpa armigera* (Hubner)	Cotton, chickpea, pigeonpea, sunflower, tomato, *etc.*
Beet armyworm	*Spodoptera exigua* (Hub.)	Chickpea in southern India
Spotted pod borer	*Maruca vitrata* (Geyer)	Pigeonpea, cowpea, lab-lab beans
Diamondback moth	*Plutella xylostella* (L.)	Cabbage, cauliflower
Pink stem borer	*Sesami ainferens* (Walk.)	Maize, sorghum, wheat
Whitefly	*Bemisia tabaci* (Gen.)	Cotton, tobacco
Brown planthopper	*Nilaparvata lugens* (Stal)	Rice
Green leafhopper	*Nephotettix* spp.	Rice
Serpentine leaf miner	*Liriomyza trifolii* (Burg.)	Cotton, tomato, cucurbits, several other vegetables
Fruit fly	*Bactrocera* spp.	Fruits and vegetables
Mealy bugs	*Paracoccus marginatus* Williams & Granara de Willink	Several field and horticultural Crops
	Phenococcus solenopsis (Tinsley)	
	Cerolastodes cajaninae (Mask.)	
Thrips	Several species	Groundnut, cotton, chillies, citrus, Pomegranate
Wheat aphid	*Macrosiphum miscanthi* (Takahashi)	Wheat, barley, oats
Pod sucking bugs	*Clavigralla* spp.	Pigeonpea
Gall midge	*Orseolia oryzae* (Wood-Mason)	Rice
Termites, white grubs,	Several species	Many crops
Sugarcane aphid	*Ceratovacuna lanigera* (Zehnt.)	Sugarcane
Spotted pod borer	*Maruca vitrata* (Fabricius)	Pigeonpea and Cowpea
Fall army worm	*Spodoptera exempta* (Walk.)	Maize and Sorghum

Climate Change Effects on Geographic Distribution and Population Dynamics

Low temperatures are often more important than high temperatures in determining the geographical distribution of insect pests. Increasing temperatures may result in a greater ability to overwinter in insect species limited by low temperatures at higher latitudes, extending their geographical range (EPA, 1989; Hill and Dymock, 1989). There are several examples of changes in the geographic distribution of several insect species as a result of climate change (Table 3). However, whether or not an insect pest would move with a crop into a new habitat will depend on other environmental conditions such as the presence of overwintering sites, soil type and moisture. The cotton bollworm/legume pod borers, *H. armigera* and *M. vitrata* will move to temperate regions in northern Europe (Figure 2). *Helicoverpa armigera* has already reached Brazil as an invasive pest, and is likely to move to

Table 3: Examples of Changes in Geographic Distribution of Insect Pests as a Result of Climate Change

Insect Pest	*Host Plant/s*	*Impact on Insects/ Behavioral Response*	*Reference*
Corn earworm, *Helicoverpa zea* (Boddie)	Maize	Range expansion to higher altitudes and northern Europe and USA, and increased overwintering	Diffenbaugh *et al.* (2008)
European corn borer, *Ostrinia nubilalis* (Hub.)	Maize	Northward shifts with an additional generation per season Increased presence in southern	Porter *et al.* (1991)
Old world Bollworm *Helicoverpa armigera*		Europe and outbreaks	Cannon (1998)
Pod borers, *Helicoverpa armigera* and *Maruca vitrata* (Geyer)	Cotton, pulses, vegetables	Expansion of geographic range in northern Asia and Europe	Sharma (2013)
Oak processionary moth, *Thaumetopoea processionea* (L.)		Northward range extension from southern Europe to Belgium, Netherlands and Denmark	Cannon (1998)
Non-migratory butterflies		Pole ward shift of geographic range	Parmesan and Yohe (2003)
Diuraphis noxia	Cereals	China, EU, USA	Battistil and Larsson, 2015
Phthorimaea operculella	Solanaceous	Worldwide	Kroschel *et al.*, 2013
Mediterranean Corn Borer (MCB) *Sesamia nonagrioides*	Maize	Europe and Russia	Maiorano *et al*, 2014
Halotydeus destructor (redlegged earth mite)		Western Australia to Southern Australia	McDonald *et al.* (2015)
Zaprionus indianus (African fig fly)	Fig	Africa continent to India	Gibert *et al.* (2016) and Macfadyen *et al.* (2018)

North America (Czepak *et al.*, 2013). Fruit flies, *Bactrocera tryoni* (Froggatt), *Bactrocera cucurbitae* (Coquillett) and *Bactrocera latifrons* (Hendel), may spread into colder areas due to increasing temperature (Prabhakar *et al.*, 2012). In addition to the direct effects of temperature changes on development rates, changes in food quality due to plant stress may result in dramatic increases in growth of insect pest populations, while the growth of certain insect pests may be adversely affected (Maffei *et al.*, 2007). Some species might extend their geographic range towards the pole and to higher altitudes (Porter *et al.*, 2014; Svobodova *et al.*, 2014). Potential changes in temperature, rainfall and wind patterns associated with climate change are expected to have a dramatic effect on desert locust in Africa, the most dangerous of all migratory pests (Cressman, 2013).

Global Warming and Pest Outbreaks

Global warming will result in pest outbreaks (Table 4). Pest outbreaks are more likely to occur with stressed plant's as a result of weakening of plants' defensive system, and thus, increasing the level of susceptibility to insect pests. Global

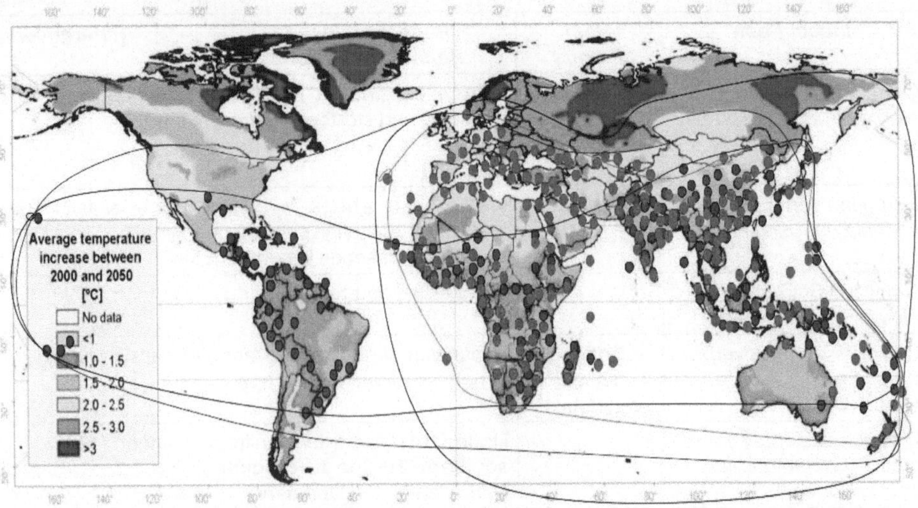

Figure 2: Likely Changes in Geographical Distribution of Cotton Bollworm/ Legume Pod Borer, *Helicoverpa armigera* (Red line: current, and black line likely distribution in future), and spotted pod borer, *Maruca vitrata* (Blue line current, and brown line future distribution) (Sharma 2013).

Table 4: Recent Pest Outbreaks in Relation to Climate Change in India

Insect Pest	*Host Plant/s*	*Regiosssn/ Location*	*Probable Reason/s*	*Impact*	*Reference*
Sugarcane woolly *aphid* *Ceratovacuna* lanigera	Sugarcane	Karnataka and Maharashtra during 2002-03	Abnormal weather Conditions Insecticide misuse	30 % yield Losses reduced cane recovery	Joshi and Viraktamath (2004), Srikanth (2007)
Plant hoppers *Nilparvata lugens* *Sogatella furcifera*	Rice	North India	Abnormal weather conditions Insecticide misuse	Crop failure on >33,000 Ha	IARI (2008)
Mealybug. *Phenacoccus solenopsis*	Cotton, Vegetables	Punjab, Haryana	Hot and dry weather Insecticide misuse	30-40% loss	Dhawan *et al.* (2007)
Papaya mealybug, *Paracoccus marginatus*	Papaya	Tamil Nadu, Karnataka, Maharashtra	Abnormal weather Conditions Misuse of insecticides	Significant yield loss	Tanwar *et al.* (2010)
South American tomato pinworm	Tomato	Karnataka and MH		Significant yield loss	IAPPS(2015)

warming will lead to earlier infestation by H. armigera in North India (Sharma 2010, 2013), resulting in increased crop loss. Many insects such as Helicoverpa spp. are migratory, and therefore, may be well adapted to exploit new opportunities by moving rapidly into new areas as a result of climate change (Sharma 2005).

Climate Change Effects on Expression of Resistance to Insect Pests

Host plant resistance to insects is one of the most environmental friendly components of pest management. However, climate change may alter the interactions between the insect pests and their host plants (Sharma 2012). Resistance to sorghum midge, observed in India, breaks down under high humidity and moderate temperatures in Kenya (Sharma *et al.*, 1999). There will be increased impact on insect pests which benefit from reduced host defenses as a result of the stress caused by the lack of adaptation to sub-optimal climatic conditions. The introduction of new crops and cultivars to take advantage of the new environmental conditions is one of the adaptive methods suggested as a possible response to climate change (Parry and Carter, 1989).

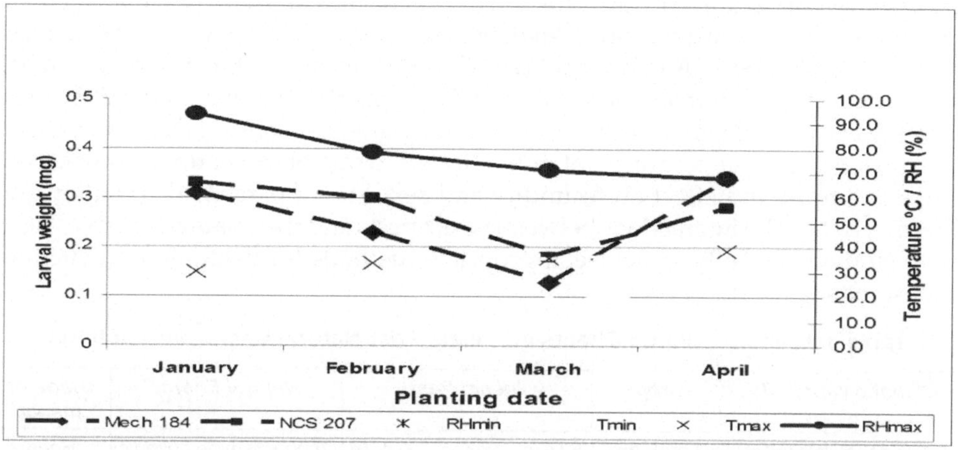

Figure 3: Effect of Climatic Factors on Expression of Resistance (Larval weight) in *Bt* Transgenic Cotton to Bollworm, *Helicoverpa armigera* (Sharma 2013).

The effects of increased atmospheric CO_2 on herbivory will not only be species-specific, but also specific to each insect-plant system. Increased CO_2 may also cause a slight decrease in nitrogen-based defenses (*e.g.*, alkaloids) and a slight increase in carbon-based defenses (*e.g.*, tannins). Lower foliar nitrogen due to CO_2 causes an increase in food consumption by the herbivores up to 40%, while unusually severe drought increases the damage by insect species such as spotted stem borer, *Chilopartellus* (Swin.) in sorghum (Sharma *et al.*, 2005).Maximum larval, pupal weights of *H. armigera* were recorded in insects reared under lower temperature and larval, pupal periods, weight and fecundity were increased with an increase in CO_2 concentration (Sharma *et al.*, 2016).

Environmental factors such as soil moisture, soil fertility, and temperature have strong influence on the expression of *Bt* toxins in transgenic plants (Sachs *et al.*, 1998; Sharma, 2013) (Figure 3). Possible causes for the failure of insect control may be: inadequate production of the toxin protein, effect of environment on transgene expression, locally resistant insect populations, and development of resistance due to inadequate management (Sharma and Ortiz, 2000). It is therefore important to understand the effects of climate change on the efficacy of transgenic plants for pest management.

Climate Change Effects on Activity and Abundance of Natural Enemies

Relationships between insect pests and their natural enemies will change as a result of global warming, resulting in both increases and decreases in the status of individual pest species (Table 5). Changes in temperature will also alter the timing of diurnal activity patterns of different groups of insects, and changes in interspecific interactions could also alter the effectiveness of natural enemies for pest management (Hill and Dymock 1989). Quantifying the effect of climate change on the activity and effectiveness of natural enemies will be a major concern in future pest management programs. Oriental army worm, *Mythimna separata* (Walk.) populations increase during extended periods of drought (which is detrimental to the natural enemies), followed by heavy rainfall (Sharma *et al.*, 2002). Aphid abundance increases with an increase in CO_2 and temperature, however, the parasitism rates remain unchanged in elevated CO_2. Temperature not only affects the rate of insect development, but also has a profound effect on fecundity and sex ratio of parasitoids (Dhillon and Sharma, 2009). The interactions between insect pests and their natural enemies need to be studied carefully to devise appropriate methods for using natural enemies in pest management.

Table 5: Climate Change Effects on Insect Pest-Natural Enemy Interactions

Climatic Variability	*Crop*	*Insect Pest*	*Natural Enemy*	*Stage of Insect*
Decreased rainfall in Sept and Oct	Sorghum	Stem borer, *Chilo partellus* (Swin.)	*Trichogramma chilonis* Ishii	Eggs
Increased rainfall variability	Castor	Semilooper, *Achaea janata* Lin.	*Trichogramma chilonis* Ishii	Eggs
Decrease in rainfall in Jun-Sept	Soybean	Leaf eating caterpillar, *Spodoptera litura*(Fab.)	*Telenomus remus* (Nixon) *Cotesia flavipes* (Cam.)	Eggs Larvae
Increase in rainfall events	Groundnut	Leafminer, *Aproaerema modicella* (Dev.)	*T. chilonis* *T. remus*	Eggs
		S. litura	*C. flavipes*	Larvae
Dry weather conditions	Chickpea, pigeonpea	Pod borer, *Helicoverpa armigera* (Hub.)	*Campoletis chloridae* Uchida	Larvae
Decrease in August rainfall	Rice	Yellow stem borer, *Scripophaga incertulas* (Walk.)	*Tetrastichus* spp.	Eggs

Source: Rao et al., (2010).

Table 6: Advances in Crop Insect Modeling Methods-towards a Whole System Approach

Sl. No.	*Forecast Models*	*Applications*	*Crop Pest/Diseases*	*Location*	*References*
1	Matrix models	Over several decades, for describing the changes that occur on the population densities of insects in a region, matrix models have been widely used	*Paronychiurus kimi*	South Korea	Lewis, 1977
2	Phenology models	Several authors have shown that the physiological response of most poikilothermic species is adapted to a particular temperature range, which is often considered as an abiotic factor influencing the species abundance and establishment in a region	Stem borers, *Busseola fusca* (Fuller), *Sesamia calamistis* Hampson, *Phenacoccus solenopsis*	India, Kenya	Briere *et al.*, 1999; Fand *et al.*, 2014a, 2014b.
3	Differential equations	Several differential equations in continuous and discrete forms are used to explain empirical datasets for single species pest population fluctuations over time	*Plutella xylostella*	Kenya	Nedorezov *et al.*, 2008; Tonnang et al.,2009
4	Competition model	Systems of differential equations are developed to explore the effects of resource (crop) and competition between insect pestspecies	Sap feeders	USA	Kaplan and Denno, 2007.
5	Fitting models	In fitting models, gradient matching method or the develop-ment of a loss function that uses the difference between predictions of the model equations and empirical time-series of the population density obtained from field surveys is usually derived	DBM *Plutella xylostella*	Russia	Ellner *et al.*, 2002.
6	GENEPEST model	On the American Phytopathology Society website (APSnet), an educational module on Simulation Modelling in Botanical Epidemiology and Crop Loss Analysis provides an overview of PDM and crop loss models. A simple crop growth simulation model for actual growth and yield, and yield losses	Borers, aphids, rust diseases	USA	Savary *et al.*, 2006,
8	NAPPFAST	North-CarolinaState-University/Animal-and-Plant-Health Inspection-Service Plant Pest Forecasting System. nappfast is particularly designed to support the requirements of the US Department of Agriculture, uses customized climatic data sets and is thus difficult for other countries to apply.	*Bactrocera dorsalis*	Italy	Donatelli *et al.*, 2017
9	PRATIQUE DSSs	For the EU, for mapping climatic suitability the area of potential establishment and the area at highest risk	Western corn rootworm, *Diabrotica virgifera* virgifera	Netherlands	Eyre *et al.*, 2012, Baker *et al.*, 2012.

Sl. No.	Forecast Models	Applications	Crop Pest/Diseases	Location	References
10	BioMA-Diseases	This modeling framework is composed by four extensible software libraries targeting the modeling of a generic fungal plant diseases. It provides input/output data structures and models to simulate a polycyclic fungal plant epidemic and to quantify its impact on crop growth.	Most of fungal diseases		Bregaglio and Donatelli, 2015
11	CIPRA	Computer Centre for Agricultural Pest Forecasting –Canada	Major insect pests	Canada	Dominique Plouffe *et al*,2014
12	SOPRA	Developed with the objective of optimizing timing of monitoring, management and control measures of insect pests in fruit orchards in	*Cydia pomonella*, *Grapholita lobarzewskii*,	Switzerland.	
13	MORPH pest models	AHDB Horticulture (formerly HDC) is the only major funder of applied horticultural research in the UK.	Cabbage root fly, carrot fly, pollen beetle, large narcissus fly	UK	Samietz *et al* 2007
14	ARIMA	Autoregressive Integrated Moving Average (ARIMA) model of time-series analysis-	Stink bug, *Tessaratoma papillosa*	Meghalaya	Boopathi *et al* 2015
15	Climax and Dymex	CLIMEX is a bioclimatic niche model that has been demonstrated to be well-suited to estimating the potential distribution of a wide range of taxa. The CLIMEX Compare Locations model	*Helicoverpa armigera*, root lesion nematodes (*Pratylenchus thornei*)	Australia	Kriticos *et al*, 2015
16	ILCYM	The ILCYM software is a tool that assists in developingphenology models, conducts population analysis and risk mappingusing this principle	*Helicoverpa armigera*, *S. litura*	India and Africa	Fand *et al.*, 2014a, Tonnanga *et al* 2017
17	Maxent	maxent is part of a suite of models termed alternatively species distribution models, habitat suitability models, niche models, environmental niche models, bioclimatic models and climate envelopes, among others	Turtle, cut worm *etc.*	France	Phillips *et al.*, 2006; Yoan Fourcade *et al*, 2014, Phillips and Dudík, 2008
18.	PEST MAN	PEST-MAN, has an intuitive style of presentation involving simple pull-down menus, context-sensitive help statements and colour graphics, and allows users to familiarize themselves quickly with the system	Apple and Pear pests	Canada	Morgan and Salomon, 1993
19.	SIMLEP	Developed decision support system for Colorado potato beetle	Colorado potato beetle (*Leptinotarsa decemlineata*)	Germany and Austria	Jorg *et al*,2007

Finally, we need to adopt and assess the efficacy of various integrated pest and disease management systems that take into consideration the change in pest spectrum, cropping patterns and effectiveness of different components of pest management for sustainable crop production (Table 6).

References

Battisti, A. and Larsson, S., 2015. Climate Change and Insect Pest Distribution Range Ed *Climate Change and Insect Pests*. CABI, 1-15.

Cannon, R.J.C., 1998. The implications of predicted climate change for insect pests in the UK, with emphasis on non-indigenous species. *Global Change Biology.*, 4 : 785-796.

Chakraborty, S. and Newton, A.C., 2011. Climate change, plant diseases and food security: an overview. *Plant Pathology*, 60: 2-14.

Cressman, K., 2013. Climate change and locusts in the WANA Region. *In* M.V.K Sivakumar, R. Lal, R. Selvaraju & I. Hamdan, eds. *Climate change and food security in West Asia and North Africa*, pp. 131–143. Springer. DOI 10.1007/978-94-007-6751-5_7.

Czepak C, Albernaz, K. C., Vivan, L. M., Guimarães HO, Carvalhais T. 2013. Primeiroregistro de ocorrência de *Helicoverpa armigera* (Hübner) (Lepidoptera: Noctuidae) no Brasil. *Pesquisa Agropecuária Tropical* 43: 110-113.

Baldocchi, D. and Valentini, R., 2004. Geographic and temporalvariation of the mechanisms controlling carbon exchange by ecosystems and their sensitivity to environmental perturbations. In: Field CB, Raupech M, eds. SCOPE 62. The Global Carbon Cycle: Integrating Humans, Climate, and the NaturalWorld. Washington, DC, USA: Island Press, 295-315.

Barnett. C, Hossell, J., Perry, M., Procter, C. and Hughes, G., 2006. A Handbook of Climate Trends Across Scotland. Scotland: Scotland and Northern Ireland Forum for Environmental research (SNIFFER): SNIFFER project CC03.

Churkina, G. and Running, S.W., 1998. Contrasting climatic controls onthe estimated productivity of global terrestrial biomes. Ecosystems 1: 206-15.

Crowley, T. J., 2000. Causes of climate change over the past 1000 years. *Science* 289: 270-277.

Dhaliwal, G. S., Jindal, V. and Dhawan, A.K., 2010. Insect pest problems and crop losses: Changing trends. *Indian Journal of Ecology*, 37: 1-7.

Dhawan, A. K., Singh, K., Saini, S., Mohindru, B., Kaur, A., Singh, G. and Singh, S., 2007. Incidence and damage potential of mealybug, *Phenacoccus solenopsis* Tinsley, on cotton in Punjab. *Indian Journal of Ecology* 34 : 110-116.

Dhillon, M. K. and Sharma, H.C., 2009. Temperature influences the performance and effectiveness of field and laboratory strains of the ichneumonid parasitoid, *Campoletis chlorideae*. *Bio Control*, l54 : 743-750.

Dinesh, D., Bett, B., Boone, R. and Grace, De 2015, Impact of climate change on African agriculture: focus on pests and diseases. CGIAR report. 1-4.

Diffenbaugh, N. S., Krupke, C. H., White MA and Alexander CE. 2008. Global warming presents new challenges for maize pest management. *Environment Research Letter* 3 : 1-9.

Environment Protection Agency. 1989. The potential Effects of Global Climate Change on the United States. Vol 2: National Studies. Review of the Report to Congress, US Environment Protection Agency, Washington DC, USA.

FAO. 2016. Climate change and food security: risks and responses ISBN 978-92-5-108998-9

FAO 2003. Trade Reforms and Food Security. Conceptualizing the Linkages. Rome, Italy: Food and Agriculture Organisation of the United Nations. http://www.fao. Org/docrep/005/y4671e/y4671e00.htm. Accessed 12 Nov 2010.

FAO., 2015. *International Standards for Phytosanitary Measures, ISPM 5, Glossary of Phytosanitary Terms*. International Plant Protection Convention (available at https://www.ippc.int/static/media/files/publication/en/2015/05/ISPM_05_En_2015-05-29_CPM-10.pdf).

Fand BB, Kamble AL and Kumar M. 2012. Will climate change pose serious threat to crop pest management: A critical review? *International Journal of Scientific Research Publications* 2 : 2250-3153.

Hill MG and Dymock JJ. 1989. Impact of Climate Change: Agricultural/Horticultural Systems. DSIR Entomology Division Submission to the New Zealand Climate Change Program. Auckland, New Zealand: Department of Scientific and Industrial Research.

Houghton RA. 1995 Effects of land-use change, surface temperature, and CO_2 concentration on terrestrial stores of carbon. In: Biotic Feedbacks in the Global Climate System [Woodwell, G.M. (ed.)]. Oxford University Press, London, pp. 333-350

Indian Agricultural Research Institute News. 2008. Brown plant hopper outbreak in rice. IARI News 24 : 1-2.

IPCC 1990.The Potential Impacts of Climate Change on Agriculture and Forestry. Intergovernmental Panel on Climate Change. Geneva and Nairobi, Kenya: World Meteorological Organization and UN Environment Program.

IPCC. 2014.*Climate change 2014: synthesis report*. Contribution of Working Groups I, II and III to the Fifth Assessment Report of the Intergovernmental Panel on Climate Change. Core Writing Team, R.K. Pachauri & L.A. Meyer, eds. Geneva, Switzerland, IPCC. 151 p.

Joshi S and Viraktamath CA. 2004. The sugarcane woolly aphid, *Ceratovacuna lanigera* Zehntner (Hemiptera: Aphididae): its biology, pest status and control. *Current Science* 87 : 307-316.

Kirtman B, Power SB, Adedoyin JA, Boer GJ, Bojariu R, Camilloni I, Doblas-Reyes FJ, Fiore AM, Kimoto M, Meehl GA, Prather M, Sarr, A, Schär C, Sutton R, van Oldenborgh GJ, Vecchi G. and Wang, H.J. 2013. Near-term climate change: projections and predictability. *In* T.F. Stocker, D. Qin, G.-K. Plattner, M. Tignor,

S.K. Allen, J., Boschung, A. Nauels, Y. Xia, V. Bex& P.M. Midgley, eds. *Climate change 2013: the physical science basis.* Contribution of Working Group I to the Fifth Assessment Report of the Intergovernmental Panel on Climate Change, Cambridge, UK, and New York, USA, Cambridge University Press

Maffei ME, Mithofer A and Boland W. 2007. Insects feeding on plants: Rapid signals and responses proceeding induction ofphytochemical release. *Phytochemistry* 68 : 2946-2959.

NOAA.2016. ESRL Global Monitoring Division. (2016, April 5). Retrieved April 25, 2016, from http://www.esrl.noaa.gov/gmd/

Oerke EC. 2006. Crop losses to pests. Journal of Agricultural Science 144: 31–43.

Parmesan C and Yohe G. 2003. A globally coherent fingerprint of climate change impacts across natural systems. *Nature* 421: 37- 42.

Parry ML and Carter TR. 1989. An assessment of the effects of climatic change on agriculture. Climate Change 15 : 95-116.

Porter JH, Parry M L and Carter T R 1991. The potential effects of climate change on agricultural insect pests. *Agricultural and Forest Meteorology* 57 : 221-240.

Porter JR, Xie L, Challinor AJ, Cochrane K, Howden SM, Iqbal MM, Lobell DBandTravasso MI. 2014. Food security and food production systems. In C.B. Field, V.R. Barros, D.J. Dokken, K.J. Mach, M.D. Mastrandrea, T.E. Bilir, M. Chatterjee, K.L. Ebi, Y.O. Estrada, R.C. Genova, B. Girma, E.S. Kissel, A.N. Levy, S. MacCracken, P.R. Mastrandrea& L.L. White, eds. Climate change 2014: impacts, adaptation, and vulnerability. Part A: global and sectoral aspects, pp. 485–533. Contribution of Working Group II to the Fifth Assessment Report of the Intergovernmental Panel on Climate Change. Cambridge, UK, and New York, USA, Cambridge University Press.

Prasad Y G and Bambawale OM. 2010. Effects of Climate Change on Natural Control of Insect Pests.*Indian Journal of Dryland Agriculture Research and Development* 25 : 1-12.

Prabhakar CS, Sood P, Mehta PK. 2012. Fruit fly(Diptera: Tephritidae) diversity in cucurbit fields andsurrounding forest areas of Himachal Pradesh, a North Western Himalayan state of India. *Arch. Phytopathol. Plant Protect.* 45: 1210–1217.

Rao GGSN, Rao VUM, Vijaya Kumar P and Rao AVMS.2010. 25 Years Research of AICRP on Agrometeorology.Central Research Institute for Dryland Agriculture, Santhoshnagar, Hyderabad, Telengana, India. 112p.

Sachs ES, Benedict J H, Stelly DM, Taylor JF, Altman DW, Berberich SA and Davis S K. 1998. Expression and segregation of genes encoding Cry1A insecticidal proteins in cotton. *Crop Science* 38: 1-11.

Scholes RJ and Biggs R. 2004. Ecosystem Services in Southern Africa: A Regional Assessment. Pretoria, South Africa: Council for Scientific and Industrial Research.

Sharma HC. 2005. *Heliothis/Helicoverpa* Management: Emerging Trends and Strategies for Future Research. Oxford & IBH, and Science Publishers, New Delhi, India.

Sharma HC. 2010. Effect of climate change on IPM in grain legumes. In: 5thInternational Food Legumes Research Conference (IFLRC V), and the 7th European Conference on Grain Legumes (AEP VII), 26-30th April 2010, Anatalaya, Turkey.

Sharma HC. 2012. Effect of global warming on insect-host plant-environment interactions. In: 24th International Congress of Entomology, 19-24 Aug 2012, Daegu, South Korea.

Sharma, H.C. 2013. Climate change effects on activity and abundance of insects: Implications for Crop Protection and Food Security. In: Combating Climate Change: An Agricultural Perspective (Kang, M.S., and Banga, S.S, eds.). Taylor and Francis, Boca Raton, Florida, USA. pp. 213-236.

Sharma HC. 2014. Climate change effects on insects: Implications for crop protection and food security. *Journal of Crop Improvement* 28 : 229-259.

Sharma HC, Dhillon MK, Kibuka J and Mukuru SZ. 2005.Plant defense responses to sorghum spotted stem borer, *Chilopartellus* under irrigated and drought conditions. *International Sorghum and Millets Newsletter* 46 : 49-52.

Sharma HC, Mukuru SZ, Manyasa E and Were J. 1999. Breakdown of resistance to sorghum midge, *Stenodiplosis sorghicola.Euphytica*109 : 131-140.

Sharma HC and Ortiz R 2000. Transgenics, pest management and the environment. *Current Science*79 : 421-437.

Sharma HC, Sullivan DJ and Bhatnagar VS. 2002.Population dynamics of the Oriental armyworm, *Mythimnaseparata* (Walker) (Lepidoptera: Noctuidae) in South-Central India. *Crop Protection*21: 721-732.

Sharma, H C Jaba Jagdish and Kanzum Bampy.2016.Temperature and CO affect survival and development of *Helicoverpa armigera*. Conference on National Priorities in Plant Health Management February 4-5, 2016, Tirupati.

Srikanth J 2007. World and Indian scenario of sugarcane woolly aphid. In: Woolly Aphid Management in Sugarcane (MukunthanN, Srikanth J, Singaravelu B, RajulaShanthy T, Thiagarajan R and PuthiraPrathap D, eds.). Extension Publication, Sugarcane Breeding Institute, Coimbatore, Tamil Nadu, India. Pp1-12.

Svobodová E, Trnka M, Dubrovský M, Semerádová D, Eitzinger J, Stěpánek Pand Zalud Z. 2014. Determination of areas with the most significant shift in persistence of pests in Europe under climate change. *Pest Manag. Sci.*, 70(5): 708–15. doi: 10.1002/ps.3622.

Tanwar RK, Jeyakumar P and Vennila S. 2010. Papaya mealybug and its management strategies. Technical Bulletin 22.National Centre for Integrated Pest Management, New Delhi-110 012, India.

UNFCCC. 2015. *Synthesis report on the aggregate effect of the intended nationally determined contributions*. Note by the Secretariat, 30 October 2015. FCC.

14

Impact of Climate Change on Insect Pests

M. Srinivasa Rao

Introduction

There is now adequate evidence about the impending climate change and the consequences thereof. The fourth assessment report of IPCC observed that 'warming of climate system is now unequivocal, as is now evident from observations of increase in global average air and ocean temperatures, widespread melting of snow and ice, and rising global sea level (IPCC, 2007). Though climate change is global in its occurrence and consequences, it is the developing countries like India that face more adverse consequences. Third IPCC report predicts that the global average surface temperature will increase further by 1.4 – 5.8°C by 2100 with the increased atmospheric carbon dioxide (CO_2) from 540-970 ppm over the same period. Since climate is the direct input into the agriculture production process, the agricultural sector has been a natural focus for research.

Climate change projections made up to 2100 for India indicate an overall increase in temperature by 2-4°C with no substantial change in precipitation quantity (Kavikumar, 2010). However, different regions are expected to experience differential change in the amount of rainfall that is likely to be received in the coming decades. It is projected that some parts of country will receive higher amount of rainfall. Last three decades saw a sharp rise in all India mean annual temperature (Venkateswarlu, 2010). Though most dryland crops tolerate high temperatures, rainfed crops grown during *rabi* are vulnerable to changes in minimum temperatures. Analysis of data for the period 1901-2005 by IMD suggests that annual mean temperature for the country as a whole has risen to 0.51°C over the period Rising carbon dioxide will increase the carbon-to-nitrogen balance in plants, which in turn will affect insect feeding, concentrations of defensive chemicals in plants, compensation responses by plants to insect herbivory, and competition between pest species (Coviella and Trumble, 1999).

Insects are cold-blooded organisms - the temperature of their bodies is approximately the same as that of the environment. Therefore, temperature is probably the single most important environmental factor influencing their behaviour, distribution, development, survival, and reproduction. Generally the impacts of CO_2 on insects are thought to be indirect through the changes in the host crop *i.e.*, the host mediated one. The information on influence of three major factors of climate change *i.e.*, temperature, carbon dioxide and precipitation on insects is discussed under the following heads.

Increased Temperature

Climate change resulting in increased temperature could impact crop pest insect populations in several complex ways. Although some climate change temperature effects might tend to depress insect populations, most researchers seem to agree that warmer temperatures in temperate climates will result in more types and higher populations of insects.

Increased temperatures can potentially affect insect survival, development, geographic range, and population size. Temperature can impact insect physiology and development directly or indirectly through the physiology or existence of hosts. Increased temperatures will accelerate the development of several types of insects (cabbage maggot, onion maggot, European corn borer, Colorado potato beetle), possibly resulting in more generations (and crop damage) per year. In addition to the above observations some more predictions and generalizations were made by several researchers. The documented information on impact of elevated temperature on insect pests indicated that Northward migration of insects, developmental rates and oviposition, out breaks leading to epidemics, possibilities of introduction of invasive species were found to be increased. The effectiveness of insect bio-control by fungi, reliability of economic threshold levels, Insect diversity in ecosystems, parasitism by parasitoids were found to be increased under increased temperature conditions (Bale *et al.*, 2002; Srinivasa Rao *et al.*, 2009).

It is understood from the literature that increase in the surface temperatures would allow polyvoltine species with accelerated developmental rates allowing the earlier completion of life cycle and thus resulting in additional number of generations within a season. In case of, Aphids (Yamamura, 1998), *Plutella xylostella* (Morimoto *et al.*1998) and bark beetles *Ips typographus* (Jonsson *et al.*, 2009). Additional generations of insects in temperate climates as a result of increased temperatures, necessitating more insecticide applications to maintain populations below economic damage thresholds (Rosemary Collier, 2009).

Insects that spend important parts of their life histories in the soil may be gradually affected by temperature changes as compared to those that are above ground simply because soil provides an insulating medium that will tend to buffer temperature changes more than the air (Bale *et al.*, 2002). Insect species diversity per area tends to decrease with higher latitude and altitude, meaning that rising temperatures could result in more insect species attacking more hosts in temperate climates. It is to conclude that the diversity of insect species and the intensity of their feeding have increased historically with increasing temperature.

Elevated CO_2 Levels

The atmospheric CO_2 concentrations have increased by above 20 % and elevated CO_2 effects the plant growth and range of physical and chemical characteristics of the plant/crop. These include reduction in the leaf nitrogen content, changes in the defense compounds, water content, carbohydrates and leaf thickness. Indications are that exposure to elevated CO_2 levels will increase the plant photosynthesis, growth, above ground biomass, leaf area, yield, carbon and C: N ratio. These changes can influence the food quality for herbivorous insects and was well reviewed (Hunter, 2001). These changes in the leaf quality are likely to have varied effect on the performance of insect herbivores. The information on effect of elevated CO_2 on insect pests was compiled and presented by Srinivasa Rao *et al.* (2008).

The possible impacts of elevated CO_2 (*e*CO_2) on growth and development, behavior of insect pests attracted the attention of researchers and several reviews of such studies were attempted to draw conclusions on the impact of elevated CO_2 on insect pest incidence. Most of the reviews (Coviella and Trumble, 1998; Hunter, 2001; Srinivasa Rao *et al.*, 2006) are available in the literature indicating the impact of elevated CO_2 on insect pest incidence across various agro ecosystems. These reviews were mainly qualitative summaries of the studies and the conclusions drawn are not based on any statistical or quantitative analysis.

An alternative procedure to deal with the limitations of the qualitative synthesis of studies was put forward initially by Glass (1976) and came to be known as 'Meta-analysis'. The quantification of effect of elevated CO_2 on the growth and development of insect pests through statistical synthesis of published results or meta-analysis is attempted here. One of the extensively used measures in Meta analysis is the "effect size" which integrates the results from different experiments on a given subject into an index. This effect size gives the relative magnitude of the experimental treatment.

The mean effect sizes for various insect parameters varied significantly. Among the insect primary parameters consumption (2.94) and duration of insect species (0.751) were found to be significantly positive under eCO_2 and other parameters like weight (-0.46) and population abundance (-0.05) of species were negative. Insect performance indices showed positive effect size for approximate digestibility, AD (1.281) and relative consumption rate, RCR (3.61) and negative with respect to efficiency of conversion of ingested food, ECI (-3.20), efficiency of conversion of digested food, ECD (-1.891) and relative growth rate, RGR (-1.072). Meta analysis of biochemical constituents of host plants indicated that the effect sizes were found to be negative (Nitrogen) and positive (Carbon and C: N ratio) indicating a significant variation of constituents under eCO_2 condition than ambient CO_2 condition. The implications and limitations of meta analysis were discussed.

Under elevated CO_2 conditions food consumption by the insect larvae, reproduction of aphids, predation by ladybird beetle, carbon based plant defenses were found to be increased. The reduction of insect development rates, response to alarm pheromones by aphids, parasitism, efficacy of transoceanic *B. thuringiensis*, nitrogen-based plant defenses was reported by various authors.

Experiments at Central Research Institute for Dryland Agriculture (CRIDA)

Several experiments were conducted using open top chamber (OTC) facility to study the impact of elevated CO_2 levels on insects. Three square type open top chambers (OTC) of 4x4x4 m dimensions, were constructed at CRIDA, Hyderabad, two for maintaining elevated CO_2 concentrations of 700±25 ppm CO_2 and 550±25 ppm CO_2 and one for ambient CO_2. An automatic CO_2 enrichment technology was developed by adapting software SCADA to accurately maintain the desired levels of CO_2 inside the OTCs. The concentration of CO_2 in the chambers was monitored by a non-dispersive infrared (NDIR) gas analyzer. Castor, groundnut plants were grown in the three OTCs and also in the open, outside the OTCs.

- Larval duration or time from hatching to pupation in larvae of both the species (*Achaea janata* and *Spodoptera litura*) was significantly influenced by the CO_2 condition under which castor leaves offered to them. Larval duration of both species was extended by about two days when fed with elevated CO_2 foliage. Thus, larvae fed with elevated CO_2 foliage consumed more each day and over a longer period, resulting in considerably increased ingestion. (Srinivasa Rao *et al.*, 2009).
- Significant influence of elevated CO_2 on life history parameters of S. litura on groundnut was observed. The percent variation of these parameters was significant under elevated CO_2 compared with ambient CO_2.
- The percent reduction of nitrogen content and increased percent of carbon, C: N ratio and TAE (Tannic acid equivalents) was significant in groundnut and castor foliage under elevated CO_2 in (Srinivasa Rao *et al.*, 2008).
- Under elevated CO_2 conditions the increased population of aphids, *Aphis craccivora* was increased with reduced generation time on cowpea.
- *Helicoverpa armigera* larvae consumed higher amount of chickpea foliage resulting increased larval weights under eCO_2 conditions. These larvae extended their duration by two days.

Extremes of Precipitation

Many pest species favour the warm and humid environment. Both direct and indirect effects of moisture stress on crops make them more vulnerable to be damaged by pests, especially in the early stages of plant growth. There are fewer scientific studies on the effect of precipitation on insects. Some insects are sensitive to precipitation and are killed or removed from crops by heavy rains. A decrease in winter rainfall could result in reduced aphid developmental rates because drought-stressed tillering cereals reduce the reproductive capacity of overwintering aphids (Pons and Tatchell, 1995). Some insects are killed or removed from crops by heavy rains - onion thrips (Reiners and Petzoldt 2005). Precipitation cause epidemiology of many pathogens, that depends on moisture for dispersal which in turn effects the insect population.

Impact of Climate Change on Distribution of Insect Species or Shifts

Climate plays a major role in defining the distribution limits of an insect species. With changes in climate, these limits are shifting as species expand into higher latitudes and altitudes and disappear from areas that have become climatically unsuitable. Such shifts are occurring in species whose distributions are limited by temperature such as many temperate and northern species. It is estimated that there would be shift in the crop cultivation due to shifts in the climate change which in turn may cause shifts in incidence of pests. (It is observed that a northward shift in the production of rice and maize in the northern hemisphere-major uncertainties remain in the distribution and magnitude of climate change outcomes particularly the pattern of pests.) It is well understood that the distribution of insect species is well impacted by the change in climate and unforeseen shifts in species is expected.

Prediction of Pest Scenarios

The prediction of possible pest situation in the future period based on elevated CO_2 and increase in temperature will indicate the expected possible pest scenarios. The predicted pest scenarios will improve the preparedness of the farmer to tackle the pest situation effectively. Insects (and mites) are cold blooded organisms and therefore their development rate is directly related to temperature-as temperatures get warmer insects develop faster. In addition, most insects have a lower and upper temperature development threshold, below or above which, respectively, they do not develop. Prediction of pest scenarios using different approaches *viz.*, thumb rules, simple pest models, accumulation of degree days and construction of life tables *etc.* are being attempted.

Thumb rule: The successful and standard thumb rule developed for *Helicoverpa* by NCIPM was used to predict pest incidence during future climate change periods. Thumb rule indicates that "if rainfall during month of June-sept is below the normal (500 ± 25 mm) and rainfall during the month of November is above the normal (10 ± 3 mm) then pest incidence will be severe on pulses". The future climate data was down scaled for various areas of the country using PRECIS A1B Scenario data. The future precipitation levels were estimated and used for prediction of pest scenarios. Results indicated that the predicted pest incidence would be low during 2020 and moderate during 2050 and 2080 years at Gulbarga and Hyderabad regions. (Srinivasa Rao *et al.*, 2008).

The pink bollworm incidence was estimated using prediction rules for level of pest population during 41-52 standard meteorological weeks. The thumb rule has four criteria *viz.* i. Total rainfall during 32-40 SW (Aug-Sep) is <500 mm ii. Total rainfall during 41-52 SW (Nov-Dec) is <20 mm iii. Average Max T during 41-52 SW is <33°C mm iv. Average RH during 41-52 SW is <70 %. (Case 1 if all criteria are satisfied-pest population expected to be severe, and Case 2 if two or three are satisfied Moderate and Case 3 if none or one criterion is satisfied Low.

Based on above criteria and rules the predicted and observed pest levels during 12 year period was estimated and was found that predicted and observed values were similar in all cases). These criteria were used for estimating the levels of pest population during 2020, 2071, 2086 and 2100. The results indicated that the level of population would be moderate in both 2020 and 2071; severe in 2086 and 2100 years. (Srinivasa Rao *et al.*, 2008).

Simple Pest Models: The future projected data were obtained from UK met office using Had CM3 model output under A2 a scenarios for 2020, 2050, 2071 to 2100 years. This data on precipitation, minimum temperature and maximum temperature were collected grid wise and used for prediction purpose. The data was converted to daily and standard week wise and utilized for estimating the pest scenarios using simple pest models which were developed at CRIDA. The predicted results indicated the incidences of Castor Semilooper and *Helicoverpa* would be severe during 2086 and 2100. (Srinivasa Rao *et al.*, 2008).

Degree days: Growing degree days (GDD or DD) are a measure of the "heat units" (related to temperature and the amount of time per day that an insect spends actively growing) that accumulate over time. Therefore, by keeping track of the number of DD that accumulate over time we can predict activity of different insects and their susceptible life stage(s) (target life stage for control measures). Studies were conducted to estimate the impact of increase in temperature on number of generations of tobacco caterpillar, *Spodoptera litura* on peanut for seven different locations of various agro ecological zones of the country for baseline (1961 to 1990), present (1991 to 2005), near future (2021 to 2050) and distant future (2071 to 2098) climate change (A1B) scenarios. The daily minimum and maximum temperature (MinT and MaxT) records were used to obtain cumulative degree days (DD) for each generation of insect using a temperature threshold of 10°C. Faster accumulation of degree days will making it possible for one or two additional generations with shortened life cycle (completion of generation would be 5 to 6 days earlier) of *S. litura* was inferred for both near and distant-future climate change scenarios (CCS) compared to baseline and present periods, at all locations. The additional number of generations and variation in the generation time of *S. litura* across seven growing locations of India imply the definitive and differential impacts, respectively of the projected increasing temperature in the future CCS. (Srinivasa Rao *et al.*, 2014a).

Life table approach: Investigations were conducted to understand the direct effects of rising temperature and the host mediated effects of elevated CO_2 (eCO_2) on *Spodoptera litura* (Fabricius) (Noctuidae: Lepidoptera). This study involved i. the construction of life tables of *S. litura* at six constant temperatures *viz.*, 20, 25, 27, 30, 33 and 35°C ± 0.5°C reared on peanut (*Arachis hypogaea* L.) grown under eCO_2 (550 ppm) concentration in open top chambers ii. Estimation of threshold temperatures and thermal constants and iii. Prediction of the pest scenarios during near and distant future climate change periods. Significantly lower leaf nitrogen, higher carbon and a higher relative proportion of carbon to nitrogen (C: N) were observed in peanut foliage grown under eCO_2 over ambient CO_2 (aCO_2). The mean development time (days) of each stage, egg, larva, pupa, pre-oviposition and total life span decreased from 20 to 35 °C temperature on eCO_2 foliage. The thermal requirement of *S. litura* from egg to egg (within the range of 20 °C to 35 °C) was 538.5 DD on eCO_2 as against 494.5 DD on aCO_2 foliage. Finite (λ) and intrinsic rates of increase (rm), net reproductive rate (Ro), mean generation time (T) and doubling time (DT) of *S. litura* varied significantly with temperature and CO_2 and were found to have quadratic relationships with temperature. The present results on life table parameters estimated by the bootstrap technique showed that the 'rm' values of *S. litura* on eCO_2 foliage were higher than those in the literature indicating a significant influence of eCO_2. The reduction of 'T' was noticed from a maximum of 50 days at 20 °C to minimum of 22 days at 35°C and 'λ' which is the indicator of reproductive value of new eggs was highest at 35°C and showed a negative relationship with temperature. The data on these life table parameters were plotted against temperature and two non-linear models developed for both CO_2 conditions and used for predicting the pest scenarios. Prediction of pest scenarios based on PRECIS A1B emission scenario data at eleven peanut growing locations of the country during near future (NF) and distant future (DF) climate change periods showed an increase of 'rm' and 'λ' with varied 'Ro' and reduced 'T'. The present results indicate that temperature and CO_2 are vital in influencing the growth and life table parameters of *S. litura* and that pest incidence is likely to be higher in the future. (Srinivasa Rao *et al.*, 2014b).

References

Bale Jeffery S, Gregory J Masters, Ian D Hodkinson, Caroline Awmack, T. Martijn Bezemer, Valeriek. Brown, Jennifer Butterfield, Alan Buse, John C. Coulson, John Farrar, John E. G. Good, Richard Harrington, Susane Hartley, T. Hefin Jones, Richard L. Lindroth, Malcolm C. Press, Ilias Symrnioudis, Allan D. Watt and John B. Whittaker, 2002. Herbivory in global climate change research: Direct effects of rising temperature on insect herbivores, *Global Change Biology*, 8: 1-16.

Brenda B Lin 2011 Resilience in agriculture through crop diversification: Adaptive management for environmental change. *Bio Science* 61(3): 183-193.

Coviella C and Trumble J 1999 Effects of elevated atmospheric carbon dioxide on insect plant interactions. *Conservation Biology* 13: 700-712.

Hunter M D 2001 Effects of elevated atmospheric carbon dioxide on insect-plant interactions. *Agricultural Forest Entomology* 3: 153-159.

IPCC 2007: Summary for Policymakers. In: Climate Change 2007: The Physical Science Basis. Contribution of Working Group I to the Fourth Assessment Report of the Intergovernmental Panel on Climate Change [Solomon, S., D. Qin, M. Manning, Z. Chen, M. Marquis, K.B. Averyt, M.Tignor and H.L. Miller (eds.)]. Cambridge University Press, Cambridge, United Kingdom and New York, NY, USA.

Jonsson A M, Appelberg G, Harding S and Barring L 2009. Spatio-temporal impact of climate change on the activity and voltinism of the spruce bark beetle, *Ips typographus*. *Global Change Biology*, 15: 486-499.

Kavikumar K S 2010. Climate Sensitivity of Indian Agriculture: Role of Technological Development and Information Diffusion, In: Lead Papers. 2010. National

Symposium on Climate Change and Rainfed Agriculture, February 18-20, 2010. Indian Society of Dryland Agriculture, Central Research Institute for Dryland Agriculture, Hyderabad, India. p. 192.

Morimoto N, Imura O, Kiura T 1998. Potential effects of global warming on the occurenceof Japanese pest insects. *Applied Entomology and Zoology* 33 (1): 147-155.

Pons X, Tatchell G M 1995 Drought stress and cereal aphid performance. *Annals of Applied Biology* 126: 19-31.

Reiners, S and Petzoldt C (eds.) 2005 Integrated Crop and Pest Management Guidelines for Commercial Vegetable Production. Cornell Cooperative Extension publication 124VG.

Risch Stephen J, David Andow and Miguel A Altiri 1983 Agro ecosystem diversity and pest control: Data, tentative conclusions and new research directions. *Environmental Entomology* 12: 625-629.

Rosemary Collier 2009 The impact of climate change on pests of horticultural crops. Series.p://www2.warwick.ac.uk/fac/sci/whri/research/climatechange/cgpests/

Srinivasa Rao M, Srinivas K, Vanaja M, Rao G G S N and Venkateswarlu B 2008 Impact of elevated CO_2 on insect herbivore –host interactions. *Research Bulletin*. Central Research Institute for Dryland Agriculture (CRIDA), Hyderabad, India. pp 36.

Srinivasa Rao M, Srinivas K, Vanaja M, Rao G G S N, Venkateswarlu B and Ramakrishna Y S 2009 Host plant (*Ricinus communis* Linn) mediated effects of elevated CO_2 on growth performance of two insect folivores *Current Science*. 97: 1047-1054.

Srinivasa Rao M, Rama Rao C. A., Vennila S., Manimanjari D., Maheswari M., and Venkateswarlu B. (2014a). Estimation of number of generations of *Spodoptera litura* Fab. on peanut in India during near and distant future climate change scenarios. *Scientific Research and Essays*, 9(7): 195-203.

Srinivasa Rao M, Manimanjari D, Rama Rao CA, Swathi P, and Maheswari M. (2014b). Effect of climate change on *Spodoptera litura* Fab. on peanut: a life table approach.*Crop Protection*, 66: 98-106.

Yamamura K, and Kiritani K 1998 A simple method to estimate the potential increase in the number of generations under global warming in temperate zones. *Applied Entomology and Zoology*, 33: 289-298.

Venkateswarlu, B 2009 Climate Change and Rainfed Agriculture: Research and development priorities. Key note address delivered in the International Conference on Nurturing Arid Zones for People and The Environment: Issues and Agenda for the 21st Century held at CAZRI, Jodhpur from November 24-28.

15

e-Sap: an IT Solution for Real Time Crop Health Management

A. Prabhuraj, Jyothi Patil, G.P. Uma, Shrikant, Raghavendra, Raja and H. Mallikarjun

Introduction

Pests, basically herbivorous insects and microbial diseases, have always plagued agriculture; the numbers of problems having multiplied following climate change in recent decades due to modernization. After water, they are perhaps the most worrisome to a farmer in India, and post sowing, take a major chunk of his finances. Total crop losses, even suicides by farmers, have been attributed to pests. In spite of development of several effective solutions by academias like the Universities and research institutions, private enterprises and traditional knowledge on these biotic stresses, the adoptability by the farmer has always met with poor compliance mainly due to ineffective penetration. In addition to this weak knowledge about the nature of biotic stress leading to poor identification in the field, not only by the farmer but also the extension personnel of the line department, resulted in the poor and untimely management.

Attempts to develop linkages can be followed from the origins of print media, through broadcasts over radio and television, up to short message services on mobile phones. Government and private bodies have been investing in huge manpower and resources for achieving efficient linkage. Yet, the results are far from satisfactory. Broadcasts over media of mass communication have suffered from inability to address the entire range of challenges at one go. They do not have any backward loop for information flow. Approaches like the *Krishi Vigyan Kendras, Sampark Kendras,* and others have been weighed down by the onerous task of handling the enormous populace of farmers in this Nation. With the rise in satellite-based communications, mobile telephony and the World Wide Web, there have been noteworthy efforts made to develop new kinds of linkages. However, these models also lack generality and independence. They are either dependent on a long chain of human interventions, or are unidirectional and inflexible. Delays and distortions burden them. Therefore, the real solution is in providing the field person with a dedicated tool containing relevant information that can be accessed anywhere, anytime; the information should be remotely updateable so that latest developments should be available to the field worker. The tool should allow data capture, including multimedia, from the field and communicate the same across all the players in agriculture. Real-time field scenarios should be accurately synthesized across space and time, and made available to the policymakers and researchers to facilitate decision-making. Therefore, after researching the situation, it was hypothesized that the real solution lies in providing the farmer all the information he needs right in his hands and wherever he is – to develop an *Information in Hand* model (*e*-SAP).

What is *e*-Sap?

Electronic Solutions against Agricultural Pests (*e*-SAP) model is unique in its functionality and utility. It empowers individual extension workers with *dedicated hand-held devices containing complete agriculture information*. It facilitates *bidirectional information flow of information in real time* – not only do the devices form a crucial medium for forward dissemination of useful information to farmers, but also generate return transmission of *reliable data to the scientist and policy maker*. The content is provided in the form of *high quality audio-visuals including images, animations, videos and voice*. In addition, the user will be able to flow through the content using simple logical steps to arrive at the desired information; it will not entail any search or question. Any number of *crops and their information in multilingual form can be added* to this device as a result of which, it has tremendous scope to scale up to any geographical locations.

Components of the Platform

The platform has two essential components – (i) Handheld field device: it contains the application and content; has ability for multimedia data capture; can send and receive data in GPRS/3G/Wi-Fi modes; can be operated in the farms irrespective of the availability of network; and (ii) Web-based application: it would enable retrieval and presentation of data generated from field devices; forms the entry point for agricultural content to be disseminated to the field devices; provide GIS/other graphical reports from the data generated; enable device management.

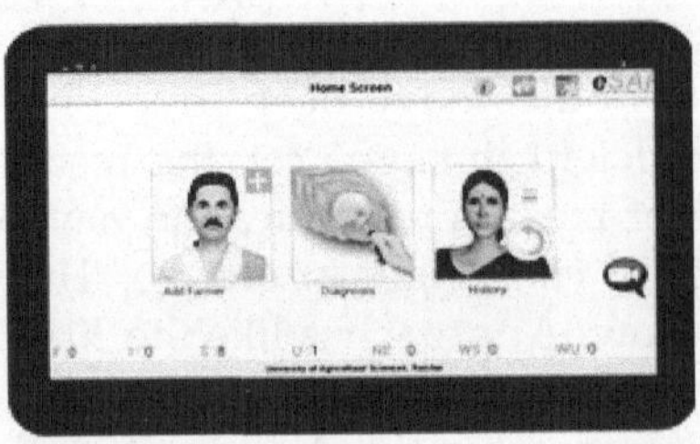

Hand held field device

Web based application

Features of *e*-sap Application

1. Pest Identification

This is one of the most highlighted features of *e*-SAP. The architecture for pest identification follows a unique image-based branching model. High-quality images that characterize pests and their symptoms are adopted to intuitively guide users in identifying the pest. Audio assistance in local language is provided at every step; the user need not be literate. The user merely needs to touch a relevant image at each of the steps to identify the problem-causing organism. The content aims at covering all known pests, so that users are able to identify all pest-related problems in the field it self; dependency on external help is minimal. As the content can be accessed offline, it can be used anywhere, anytime.

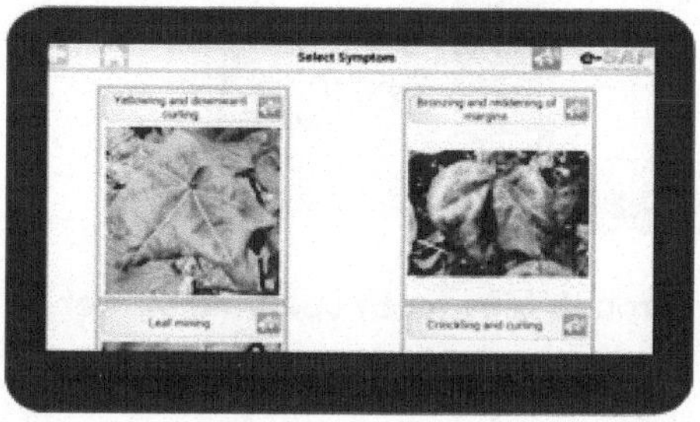

Image And Voice Assisted Symptom Architecture

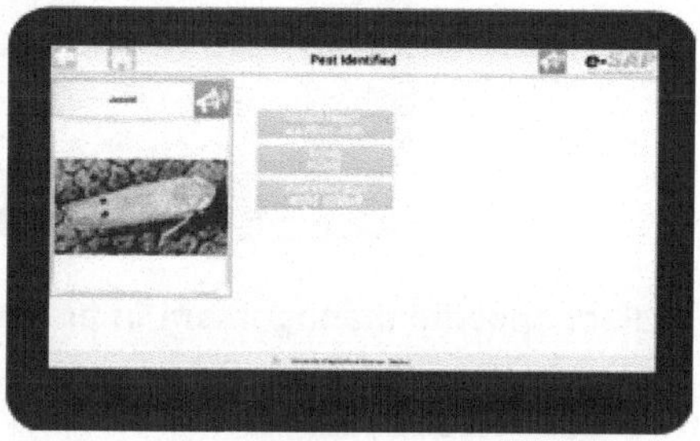

Identification of The Pest

2. Pest Surveillance

Pest identification alone is not sufficient to take up remedial measures; it is essential to determine the extent of the pest problem prevalent in each farm. For this purpose, there are intuitively built pest-specific survey forms to quantify damage caused by various pests. Data are automatically analyzed based on the survey and the per-determined economic threshold values for each pest. Results and respective suggestions are instantaneously visible on the field device. Depending on this, the user can decide on adopting management strategies or might simply watch for further buildup of the pest. Survey can be conducted offline too and results can be obtained straight away.

As surveillance entails multiple image capture by the field device, a set of close-ups and field images along with data on crop, crop age, pest damage and geo-coordinates of the field are transmitted to the cloud for further use by researchers/policymakers. Data transfer occurs instantly on the availability of telecommunication signals.

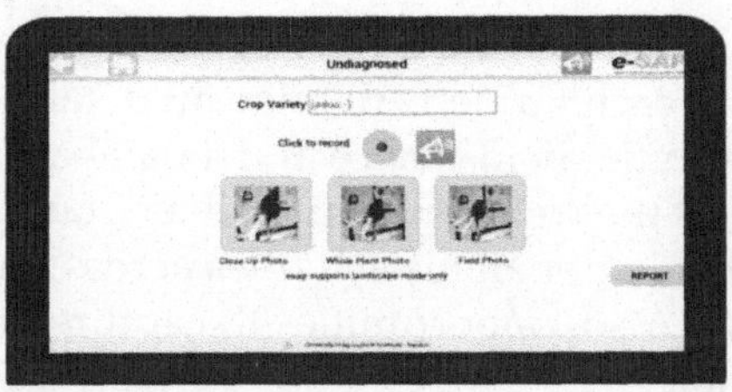

Pest/disease/deficiency specific survey modules

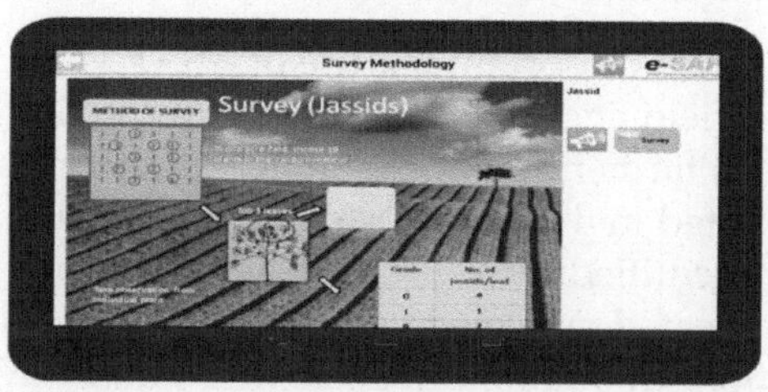

Multiple image capturing facility

3. Pest Management

A schedule of recommended management strategies are made available against each pest after determining the extent of damage. The strategy takes into account the crop, crop age and crop part affected. The user may adopt strategies depending on the automated suggestion made on the basis of the survey conducted. Management strategies are also available offline. Any new strategy, or pest management technology, can be remotely made available on the availability of telecommunication signals.

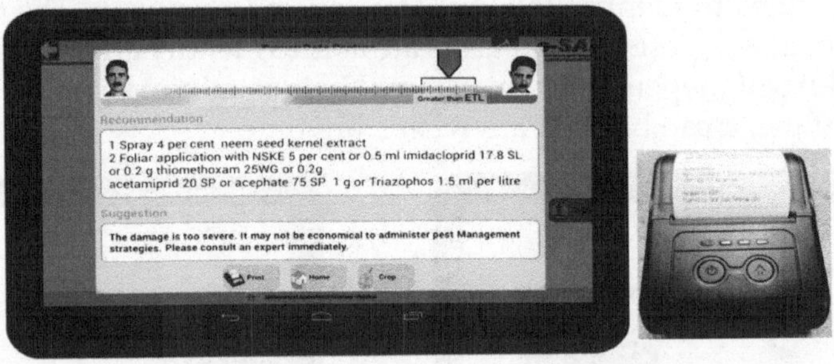

Pest specific management in printed form through blue tooth operated printer

4. Pest Information

To supplement the knowledge of users, details of each pest are made available on the field device, which is available offline and updated online.

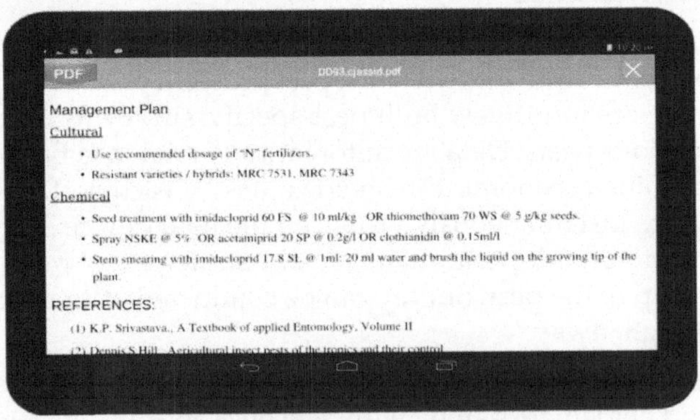

More information of the pest

5. Farmer Database

An important feature of *e*-SAP is the ability to capture farmer-specific data in the field and build database of the activities of each farmer. All relevant details of each farmer, including images, are captured on the field device and a database is created in the cloud, which is accessed through the web application. Every farmer is identified by a unique number with which a log of all his activities across time is created and made available for further use. Also, opportunities to capture any information on farm activity are made available.

Farmer registration along with his mobile phone number

6. Expert Connect

Under extraordinary cases, when the available content in the field device is insufficient or the user faces difficulty in using it correctly, expert connect is made available on the device. Here, the application not only enables capture of multiple images of the crop, but also makes it possible for the user to record his opinion as he speaks. These images and audio files along with other relevant details like ego-coordinates, crop and farmer details are transferred to the cloud in real-time. Designated experts receive an alert regarding the same, and can access information using e-SAP's web application. The application also allows inter-expert exchange of information before posting their suggestions to the field device. Normally, this process takes less than ten minutes when telecommunication networks are available to all the users.

Undiagnosed malady (UDM)

Diagnosis and resolving the UDM's by expert through e-SAP app

7. Data Analyses and Reporting

Data captured from various field devices are fed into several databases in the cloud, which are then made available for viewing over the GIS across any defined time and any chosen set of parameters.

Reports available in the form of GIS maps with all relevant field information

8. Decision Support System

Users can access automated graphs/tables over e-SAP's web application. Points over GIS maps and continuously updated graphs/tables allow real-time monitoring of pest situations across any defined space. Micro-level decision support system is provided on the field device for taking decisions on adopting pest management strategies. e-SAP also makes provision for macro-level decision support in the web application.

Based on the data generated across space and time, and built-in analyses, alerts for various scenarios can be customized by each web-user, which would warn in several ways – on mobile phones, emails, *etc.* for taking immediate actions on a larger scale.

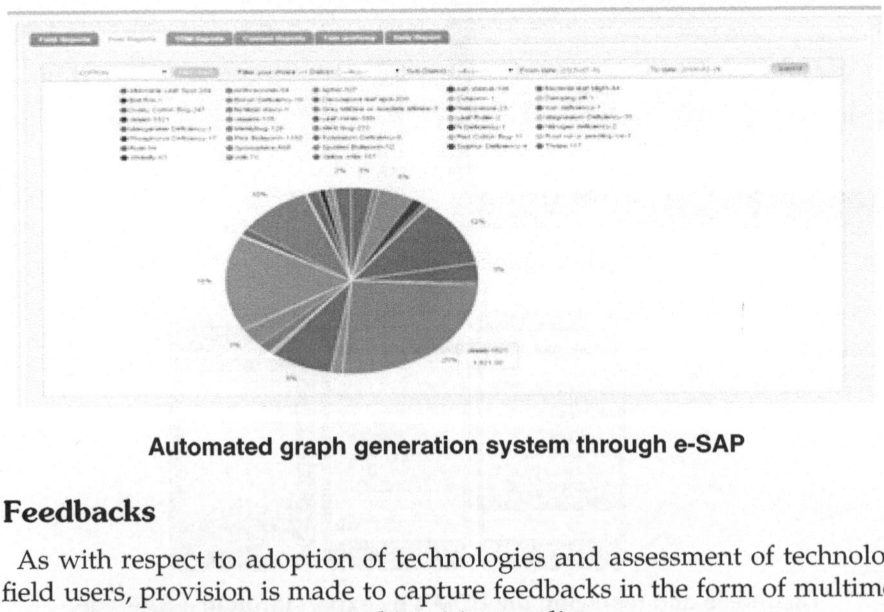

Automated graph generation system through e-SAP

9. Feedbacks

As with respect to adoption of technologies and assessment of technologies by field users, provision is made to capture feedbacks in the form of multimedia content, like audio and images, and intuitive grading.

10. Content Management

e-SAP allows for real-time dissemination of pest management technologies and other information to all/designated field devices. Information on new pests or new information on existing pests, new/modified pest identification routes, additional/new symptoms, new surveys plans, and new management strategies can be remotely updated on the field devices over existing telecommunication networks.

Content Management System

11. Device Management

The device part of the platform exploits emerging technologies with capabilities to generate/provide data in a format that an illiterate user can generate/use easy-to-understand multimedia content. Today, telecommunication networks reach a wide spectrum of geographies. However, in the absence of such networks, the device utilizes offline storage and delivers content on network availability.

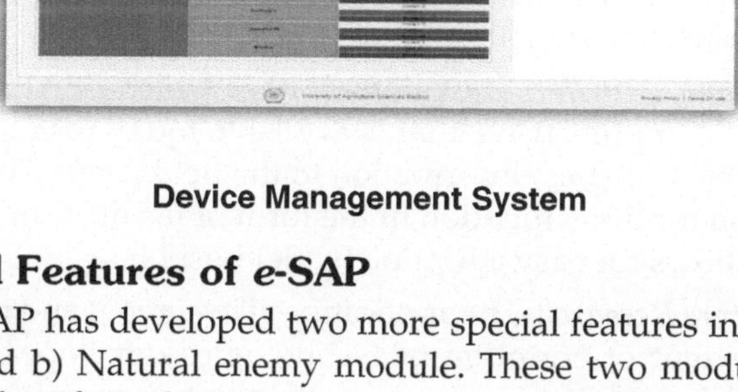

Device Management System

12. Additional Features of *e*-SAP

Recently *e*-SAP has developed two more special features in its application a) weed module and b) Natural enemy module. These two modules/components makes users to identify and management plans to specific weed problem using

"Weed module". On the other hand by using *"Natural enemy module"* one can identify and quantify the natural enemy present in the crop ecosystem.

Natural enemy module

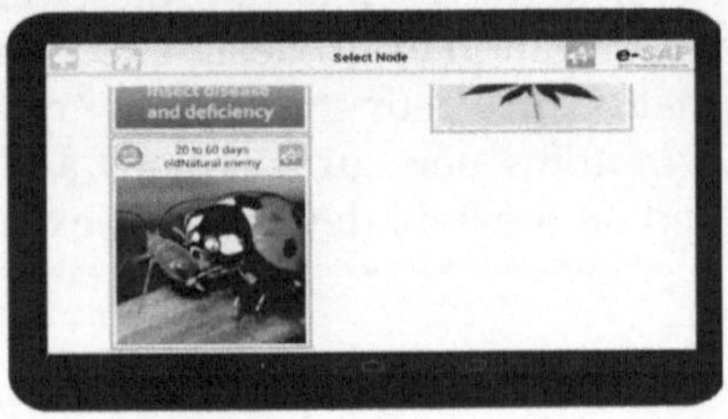

Weed survey module

Unique Features of *e*-SAP

- ☆ Empowers extension personnel in identification of pest/disease problems without the aid of entomologist/pathologist
- ☆ Enables instant solutions to identified problem right in the farmers' field
- ☆ Access to send the un-resolved problems to the expert through GPS tagged image and voice
- ☆ Option to send the GPS tagged pest/disease specific survey reports to central server
- ☆ Enables fast and deeper penetration of the recommended technologies to farmers
- ☆ Opportunity to initiate research programmes on the burning problems of the region/locality
- ☆ Opportunity to take quick and appropriate decisions based on the GIS reports generated out of survey on real time
- ☆ Bridging the gap between farmer and expert
- ☆ Strengthens the hands of extension personnel and increases the work efficiency.

Benefits of *e*-SAP

1. ***Farm Productivity:*** It enables reliable, timely and specific information to the field user in the way she/he understands by transcending language and literacy barriers, which will ultimately translate to perceptible increase in farm productivity.
2. ***Technology Transfer:*** The platform over which *e*-SAP is built is perhaps the most comprehensive that is available today for rapid dissemination of new technologies/information to the field users. The platform makes provision for dissemination in the form of intuitive multimedia content, which allows for easy adoption by field users.
3. ***Agriculture Research:*** The application allows rapid and reliable assessment of the extent of adoption of technologies developed. Additionally, it enables researchers to propose, and funding agencies to evaluate research

problems based on current farm data. Real-time data will empower researchers to design their experiments and fine tune methodologies to address on-farm challenges.

4. ***Policy and Decision-Making:*** Streaming real-time data across space and time, along with synthesized information in the form of graphs and tables, is an invaluable tool for decision-making. Additionally, customized alerts for particular areas of interest, makes real-time decisions possible. A strong foundation will be laid for targeted policies, for example, policies specific to a crop or region.
5. ***Agricultural Economy:*** Better assessments of area under cultivation, yield, crop share, pestilence, *etc.*, will lead to improved economics of the agricultural sector.

Conclusion

e-SAP has already made a difference to more than 100,000 farmers in the state of Karnataka. Farmers can now detect the crop problem in its early stages and solve it economically and safely, while real time data empower other stakeholders to take quick and reliable decisions. It has been adopted by three other Farm Universities of the state, while the state and the central governments are reviewing a proposal to adopt eSAP.

e-SAP is completely localizable in terms of language and content such that diverse local requirements can be addressed. Despite the localization, all data are captured and reports are generated. Scientifically accurate pest-specific sampling plans, incorporation of Economic Threshold Levels and advising pest intensity-based management options make *e*-SAP a powerful and unique system. It is the first to take the concept of Integrated Pest Management to the farms.

16

Effect of Climate Change on Agriculturally Important Microorganisms

P. Jones Nirmalnath

Introduction

Enhancement in concentration of atmospheric carbon dioxide (CO_2) and resultant increasing temperature as a result of the cumulative effects of anthropogenic activities, emissions from fossil fuel combustion, energy use scenario, land clearing and urbanization have created renewed interest in understanding responses of the ecosystem to increased levels of CO_2 and temperature (IPCC, 2007). The atmospheric CO_2 concentration has been steadily rising from pre-industrial era: 280 ppm in the year 1850 to 390 ppm in 2011 (Trans *et al.*, 2011). As a consequence of this rise in CO_2, the global surface temperature is predicted to increase by 1 to 6° C by the year 2100 (IPCC, 2007), leading to enhanced drought in several areas of the world, as a result of increased evapotranspiration.

Altered environmental conditions due to climate change are likely to induce changes in plant physiology and root exudation. Elevated CO_2 in particular will lead to increased C allocation to the root zone and potentially also lead to altered composition of root exudates thereby altering the availability of chemo attractants or signal compounds, C/N ratios or nutrient availability (Haase *et al.*, 2007).

Further, elevated temperature coupled with drought might lead to change in the composition, abundance and activities of plant associated microbial communities, thus having significant impact on diversity and activities of such plant associated microbial communities (Drigo *et al.*, 2008). Consequently, microorganisms known for their beneficial effects on plant growth or health might also be impaired in terms of exhibiting their desirable properties and also their colonization capacity under certain conditions.

Effect of eCO_2 and Temperature on AM Fungi

Among the beneficial soil microorganisms, approximately 90 per cent of the plants form mycorrhizal association and 60 per cent of these plants establish symbiosis with obligate symbionts such as AMF and are known to enhance plant nutrient uptake in exchange for rhizosphere soil carbohydrate compounds. However, due to climate change effects including increasing CO_2 concentrations in the atmosphere, soil warming or drought stress has indirect effect on plant associated AMF (Auge, 2001). Several studies have revealed that elevated CO_2 increased the AMF colonization in comparison with ambient CO_2, due to increased carbon allocation to their external hyphae. However, contrasting results also have been obtained by Gavito *et al.* (2000) who indicated that certain AMF strains did not respond to altered conditions. Hence, different plant growth promoting strains need to be selected for elevated CO_2 and temperature conditions

Arbuscular mycorrhizal fungi (AMF) are likely to be stimulated by elevated CO_2, which will assist P acquisition by the host plant. In this symbiotic relationship, AMF provide their host plants with mineral nutrients, such as P, in exchange for carbohydrates supplied to the AMF (Kiers *et al.*, 2011). This two-way transfer of resources is certainly affected by elevated CO_2, because elevated CO_2 increases C allocation to the roots of the host plant (Gamper *et al.*, 2004). Studies have found that the AMF hyphal network is enlarged by elevated CO_2, resulting in nutrient absorption being significantly increased (Gamper *et al.*, 2004; Staddon *et al.*, 2004). With a meta-analysis, Treseder (2004) also found that the abundance of AMF increased relative to root length under elevated CO_2. Furthermore, shifts in active AMF species under elevated CO_2 were convincingly confirmed using stable isotope (13C) probing and subsequent real-time PCR techniques (Drigo *et al.*, 2010).The increase in symbiotic activity between AMF and plants under elevated CO_2 leads to an expectation that mycorrhizal plant species will adapt better to P-deficient soils compared with non-mycorrhizal species in the elevated CO_2 environment.

Many studies reveal that increased temperature had a positive impact on AM colonization and hyphal length. In some cases, however, no effects or negative effects of enhanced temperature on AMF were reported. AMF may respond to higher temperatures with enhanced growth and plant colonization for the majority of strains (Graham *et al.*, 1982; Fitter *et al.*, 2000). This was demonstrated for instance with strains of *Glomus intraradices, Glomus mossae* as well as with many others (Baon *et al.*, 1994). Temperature may also significantly alter the structure of the AMF hyphal network and induce a switch from more vesicles (responsible for storage) in cooler soils to more extensive extra mycorrhizal hyphal networks (indicating growth) in warmer soils (Hawkes *et al.*, 2008). This may be linked to faster carbon allocation to the rhizosphere and an increased respiration of ectomycorrhizal strains of *Cenococcum geophilum, Suillus intermedius* and *Lactarius* cf. *pubescens* can be for example reduced under increasing temperature (Malcolm *et al.*, 2008). Moreover, it has also been demonstrated that *C. geophilum* morphotypes 12 and 40 have decreased colonization in *Quercus myrsinaefolia* under elevated temperature (Kasai *et al.*, 2000).

However, in other studies, soil temperature increased the percentage of total ectomycorhizal colonization such as in the case of pinyon pine (*Pinus edulis* Engelm.), with soil type- dependent results (Swaty *et al.*, 1998). Alternatively, Clemmensen *et al.* (2006) showed that the shrub *Betula nana* (dwarf birch) of arctic tundra ecosystems did not show changes in fungal biomass or ectomycorrhiza mycelial production, but warming decreased fine root biomass relative to aboveground biomass in ectomycorrhizal plants, and thus colonization of fine roots by ectomycorrhiza increased slightly. Warming may further affect the ectomycorrhiza morphotypes colonizing the roots of host plants. Therefore, not only elevated CO_2 but also increased temperatures may induce changes in ectomycorrhiza communities.

Elevation in temperature therefore may not be directly linked to changes in endophyte communities. However, plant species diversity may be impacted by warming, potentially leading to effects on endophytic fungal communities.

Effect of eCO_2 and Temperature on PGPR

Drigo *et al.* (2009) showed that elevated CO_2 increased the dominance of *Pseudomonas* sp. which are known to include many plant growth-promoting members. Similarly, the abundance of *Rhizobium* was enhanced with white clover (Marilley *et al.*, 1999). In contrast, the proportion of HCN-producing *Pseudomonas*, considered as potential inhibitors of root parasitic fungi isolated from *Medicago coerulea* was drastically reduced under elevated CO_2 conditions (Tarnawski and Aragno, 2006). However, the proportion of siderophore producers and nitrate-dissimilating strains were found to be under elevated CO_2 condition. This is an example indicating that increased atmospheric CO_2 levels might have differential impacts on plant beneficial bacteria, which might have further implications on their application in agriculture. In the case of elevated CO_2 concentrations, different inoculant strains for application in agriculture will possibly have to be selected because of the distinct performances of plant growth-promoting strains under ambient and elevated CO_2 conditions. Kohler *et al.* (2009) recently reported that a plant growth-promoting *Pseudomonas mendocina* enhanced the growth, physiological parameters and plant biomass production of lettuce plants under elevated CO_2 conditions.

Elevated atmospheric CO_2 concentrations can enhance photosynthetic rates in plants. They can therefore act as a carbon 'fertilizer' to induce increases in net ecosystem CO_2 exchange and contribute to increases in net primary productivity (Sakurai *et al.*, 2014). Thus, elevated CO_2 is likely to stimulate the growth of many plant species (Sakurai *et al.*, 2014). However, an increase in the growth of plants will need an increased supply of essential plant nutrients. In fact, limitations in supply of nutrients such as nitrogen (N) and P may offset the positive effects of elevated CO_2 on photosynthesis, thereby constraining species growth. Thus there is need for extra N and P supply under elevated CO_2 and temperature. Significant increases in P demand by plants are likely to happen under elevated CO_2 due to the stimulation of photosynthesis, and subsequent growth responses. Elevated CO_2

alters P acquisition through changes in root morphology and increases in rooting depth. Moreover, the quantity and composition of root exudates are likely to change under elevated CO_2, due to the changes in carbon fluxes along the glycolytic pathway and the tricarboxylic acid cycle. As a consequence, these root exudates may lead to P mobilization by the chelation of P from sparingly soluble P complexes, by the alteration of the biochemical environment and by changes to microbial activity in the rhizosphere. Future research on chemical, molecular, microbiological and physiological aspects is needed to improve understanding of how elevated CO_2 might affect the use and acquisition of N and P by plants.

The response of symbiotic N_2 fixation in legumes to elevated CO_2 is crucial to C sequestration in future CO_2 environments because continued C sequestration is contingent on additional N input (Hungate *et al.*, 2003). A recent review suggests that symbiotic N_2 fixation at elevated CO_2 could be enhanced as a result of CO_2-induced increase in nodule size, nodule number per plant, or specific nitrogenase activity (Rogers *et al.*, 2009). Inubishi *et al.* (2002), reported higher BNF in submerged rice paddy under elevated CO_2 in a FACE (Free-air CO_2 enrichment) experiment.

The consequences of severe abiotic limitations, elevated temperature and CO_2 on beneficial microorganisms which are the principal drivers of nutrient cycling need strong attention. This knowledge is crucial to predicting the trajectory of terrestrial environments following global warming. Studies on analysis of soil respiration to biomass ratio, microbial metabolic activity, microbial community composition *via* PLFA (phospholipids fatty acid analysis and q PCR (quantitative real time PCR) and DNA to NA ratio, have revealed a shift in the rhizosphere microflora under elevated CO_2. Zhang *et al.*, 2014 reported significant increase in the total microbial biomass in elevated temperature versus the control plots at the 0–10 cm depth, due primarily to a significant increase in the abundance of all investigated microbial groups. Hutchins *et al.*, 2015 reported irreversibly increased nitrogen fixation in *Trichodesmium*, (a marine cyanobacterium) experimentally adapted to elevated carbon dioxide.

Elevated CO_2 directly influences the density, diversity and functioning of the rhizosphere microbial communities (Haase *et al.*, 2008). Drissner *et al.* (2007) found a 481 % increase in soil microbial biomass and 125 % increase in the Shannon index (species diversity in a community) of bacterial community structure after *Trifolium repens* L. and *Lolium perenne* L. had grown under elevated CO_2 in a FACE facility for 9 years. Similarly, microbial growth rate per unit soil mass in the rhizosphere of *Populus deltoids* was up to 58 % higher under elevated CO_2 than under ambient CO_2 (Blagodatskaya *et al.*, 2010). In addition, microbial respiration and the metabolic quotient of microbes in the rhizosphere of wheat increased significantly under elevated CO_2 (Jin *et al.*, 2014).

Elevated CO_2 is able to specifically affect the abundance of some microbial genera, which may directly facilitate P solubilization in the rhizosphere. Drigo *et al.* (2009) found that the abundance of *Pseudomonas* bacteria in the rhizosphere, increased under elevated CO_2, with active populations of *P. aeruginosa, P. fluorescens, P. trivialis and P. putida* being detected. Both *P. fluorescens* and *P. putida* are considered to be P-solubilizing microorganisms that produce metabolites that solubilize sparingly soluble inorganic P compounds to release phosphate ions (Krey *et al.*, 2013). This suggests that elevated CO_2 is likely to benefit these P-solubilizing microorganisms. However, the magnitude of this effect depends on the P compounds in soils, and the plant species, which in turn will determine the abundance of the P-solubilizing microbial species in their rhizospheres.

Some rhizosphere bacteria and endophytes have been shown to alleviate temperature stress on plants, and these strains may induce growth promotion of different crops at different climates, soils and temperatures (Bashan & Holguin, 1998). Egamberdiyeva & Hoflich (2003) reported that temperature and soil type may affect the performance of plant-beneficial bacteria. *Mycobacterium* sp. 44, a *Pseudomonas fluorescens* and a *Pantoea agglomerans* strain isolated from a semi-continental climate were found to significantly increase the root and shoot growth of winter wheat at 16 1C compared with that at 26 1C in loamy sand. However, *Mycobacterium phlei* strain MbP18 as well as *Mycoplana bullata* MpB46 (all from a semi-arid climate) performed well under both conditions, indicating genotype-specific preferences for certain environmental conditions. This should be taken into consideration for the future selection of bacterial inoculants to improve plant performance. Some strains of PGPB can grow better at a high temperature than at a low temperature and could be of special interest for application in agriculture exposed to increased temperatures. In an early study, Waldon *et al.* (1989) clearly showed that rhizobia isolated from nodules of the desert woody legume *Prosopis glandulosa* grew better at 36°C than at 26°C. These strains were physiologically, however, distinct from other strains belonging to the same species, suggesting that highly adapted strains have evolved. In addition, bacteria colonizing distinct sites may react differently to certain environmental conditions. A study with the endophyte *Burkholderia phytofirmans* strain PsJN demonstrated that a temperature increase from 10 to 30°C reduced the colonization of this strain in the tomato rhizosphere, whereas endophytic abundance was not affected (Pillay & Nowak, 1997). After successful colonization, rhizosphere as well as endophytic bacteria may alleviate temperature or drought stress on plants (Aroca & Ruiz-Lozano, 2009) by inducing a systemic response (Yang *et al.*, 2009). This demonstrates the potential role of certain strains for use in agriculture, but by also taking into account the effects of changing conditions on their colonization as well as on their plant growth promoting capacities.

Research Accomplishments at UAS, Dharwad

Investigations were carried to study the impact of elevated carbon dioxide and/ temperature on arbuscular mycorrhizal fungal community in groundnut rhizosphere under the open top chamber facility at Institute for Agricultural Research on Climate Change, University of Agricultural Sciences, Dharwad, revealed that the highest soil water stable aggregate, organic carbon, inorganic carbon, available P, K and S were recorded with elevated CO_2. However, soil pH, EC and available nitrogen did not vary significantly. The population of soil bacteria, fungi, actinomycetes, free living nitrogen fixers, fluorescent pseudomonads and P- solubilizers was significantly influenced due to elevated CO_2 compared to other altered conditions. The soil enzyme activity *viz.*, dehydrogenase, urease, cellulase, acid phosphatase, alkaline phosphatase, soil biomass carbon and nitrogen were greatly influenced by elevated

CO_2 condition. The Shannon diversity index was the highest with elevated CO_2 plus temperature (H'=3.10), while elevated CO_2 recorded (H'=2.25). The Margalef's index of species richness for AMF was the highest in elevated CO_2 plus temperature (2.66) compared to ambient condition (1.37). The most predominate AM fungal species, *Glomus mosseae* from all the five altered conditions were genetically characterized by amplifying a single fragment of 375 bp in the ITS region.

Furthermore, the functional and genetic diversity of pink pigmented facultative methylotrophs (PPFM) were confirmed by screening for the presence of *mxaF* gene by amplifying a single fragment of 555 bp in all the native isolates. Based on the carbon substrate utilization test, the native PPFM isolates have been grouped under nine species *viz., M. suomiense, M. populi, M. nodulans, M. aminovorans, M. mesophilicum, M. radiotolerans, M. fujisawaense, M. extorquens* and *Methylobacterium* sp. The studies on the functional characterization of the native isolates have revealed that all the isolates were capable of producing cytokinin and IAA. Ten isolates have shown MPS activity and seven isolates were antagonistic to *Sclerotium rolfsii and Fusarium oxysporum.* Further, five efficient isolates based on their ability to produce IAA and cytokinin were characterized by sequence analysis of the 16S rDNA. The blast results clearly indicated that the isolate PPFM12 (elevated CO_2 plus temperature) and PPFM23 (elevated CO_2) were found to be very close to *M. mesophilicum*, while, PPFM33 (elevated temperature), PPFM58 (ambient condition) and PPFM65 (reference plot) were found to be very close to *M. radiotolerans*.

Conclusion and Future Prospects

The response of elevated CO_2, but also of warming and drought, will therefore also depend on the plant and the microbial genotype. The effects of elevated CO_2 on PGPB were more variable, and no general trend in terms of increased or decreased abundance of PGPB has been observed so far. However, under conditions of increased temperature or drought, different genotypes of PGPF or PGPB, potentially also showing different functional activities, might establish in the plant environment (Drigo *et al.*, 2009). However, as a result of altered community structures, beneficial microorganisms might have to compete with a different microbial community and therefore might show an altered colonization behavior. In addition to the potential effects of climate change on microbial colonization characteristics, microbial activities may be affected (Haase *et al.*, 2008). Microbial communities showing different activities or producing altered signals may, in the long term, either result in the establishment of altered communities and/or in the elicitation of different plant responses.

Considering the fact that plant-beneficial microorganisms such as mycorrhizal fungi and nitrogen-fixing bacteria provide up to 80% N and up to 75% P and that an extremely high number of plant species are completely dependent on microbial symbionts for growth and survival (van der Heijden *et al.*, 2008), it is evident that alterations in the plant-beneficial microbial communities may ultimately influence plant diversity and the diversity and functioning of the soil microbiota.

Furthermore, the abundance or the activity of microorganisms with biocontrol activities may be altered and thereby plant pathogen populations may be affected.

Climate change will induce adaptation processes in plants and microorganisms and might require the selection of adapted plant cultivars in agriculture. However, adaptation processes or the use of cultivars not fully adapted to new environmental conditions could be supported by making use of the plant growth-promotion potential of PGPF/B. Particularly under elevated CO_2 conditions, nutrients such as N might be limiting, leading to the need for enhanced fertilizer input in agriculture. Plant growth-promoting microorganisms supporting nutrient acquisition such as AM fungi or N-fixing bacteria may thus gain increasing importance.

Index

9 789354 614255

Printed at Repro India Ltd.

Printed at Repro India Ltd.